Aquaculture in the United States

Constraints and Opportunities

Aquaculture in the United States
Constraints and Opportunities

A Report of the
Committee on Aquaculture
Board on Agriculture and Renewable Resources
Commission on Natural Resources
National Research Council

NATIONAL ACADEMY OF SCIENCES
Washington, D.C.   1978

This study was supported by the National Oceanic and Atmospheric Administration, the U.S. Department of Agriculture,
and the U.S. Fish and Wildlife Service.

Library of Congress Catalog Card Number 78-52270

International Standard Book Number 0-309-02740-3

Available from:

Printing and Publishing Office
National Academy of Sciences
2101 Constitution Avenue
Washington, D.C. 20418

Printed in the United States of America

STUDY PANELS

Production Systems

E. Wayne Shell (Chairman), Auburn University
William McNeil, Weyerhauser Company
Richard A. Neal, Agency for International Development
William Shaw, National Oceanic and Atmospheric Administration
R. Oneal Smitherman, Auburn University

Science and Technology

Harold H. Webber (Chairman), Groton Associates, Inc.
John Halver, U.S. Fish and Wildlife Service
Arlene Longwell, National Marine Fisheries Service
John H. Ryther, Woods Hole Oceanographic Institution
Carl Sindermann, National Marine Fisheries Service
Robert Stevens, U.S. Fish and Wildlife Service
Durbin Tabb, Tropical Bioindustries

Economics and Business

R. Bruce Rettig (Chairman), Oregon State University
Lee Anderson, University of Delaware
Jack Davidson, Washington State University
Jon Lindberg, Domsea Farms, Inc.
Edward W. McCoy, Auburn University
J. Richards Nelson, Long Island Oyster Farms
Richard Johnston, Oregon State University

Law and Public Administration

Charles A. Black (Chairman), Mardela Corporation
Robert B. Abel, Texas A&M University
Porter Briggs, U.S. Aquaculture Council
Allan Grant, American Farm Bureau Federation
John Glude, Seattle, Washington
Marc Hershman, University of Washington
Philip Roedel, Agency for International Development

WORKSHOP PARTICIPANTS

Keen Buss, Boalsburg, Pennsylvania
John Clark, The Conservation Foundation

Wallis Clark, University of California (Davis)
J. David Clem, Food & Drug Administration
Harold Goodwin, Bethesda, Maryland
Billy Griffin, U.S. Fish and Wildlife Service
D. Heyward Hamilton, Department of Energy
William Hannum, National Oceanic and Atmospheric
   Administration
Thomas K. Hill, University of Georgia
James W. Kahrs, Osage Catfisheries
George Krantz, University of Maryland
James Lannan, Oregon State University
Cedric Lindsay, Washington Department of Fisheries
Richard Loring, Cultured Clam Corporation
Carl E. Madewell, Tennessee Valley Authority
Ronald D. Mayo, Kramer, Chin & Mayo
Jack Milnes, South Southeastern Regional Aquaculture
   Association, Inc., Alaska
Fern Wood Mitchell, U.S. Trout Farmers Association
Joan Mitchell, National Science Foundation
Richard Neve, University of Alaska
Richard Poole, LUMMI Indian Community
Gary D. Pruder, University of Delaware
Gilbert Radonski, Sports Fishing Institute
Thomas Ripley, Tennessee Valley Authority
Gary L. Rumsey, U.S. Fish and Wildlife Service
Paul A. Sandifer, Marine Resources Research Institute
Earl Splitter, Cooperative State Research Service
Lawrence Storch, Shaw, Pittman, Potts & Trowbridge
Everett A. Tolley, Shellfish Institute of North America
Emil Usinger, Bluepoints Oyster Company, Inc.
David H. Wallace, National Oceanic & Atmospheric
   Administration
Kenneth Wolf, U.S. Fish and Wildlife Service

James W. Avault, Jr., Louisiana State University
John M. Bailey, U.S. Department of Agriculture
Robert C. Baker, Cornell University
Wayne J. Baldwin, University of Hawaii (Manoa)
Robert C. Bayer, University of Maine (Orono)
L. R. Beuchat, University of Georgia
Wayne A. Bough, University of Georgia
W. P. Breese, Oregon State University
Robert W. Brick, Texas A&M University
C. F. Bryan, Louisiana State University
D. Homer Buck, Illinois Natural History Survey
Ross V. Bulkley, Iowa State University
Charles D. Busch, Auburn University
John R. Castle, Marine Colloids, Inc.
Jim Cato, University of Florida
Jean F. Caul, Kansas State University
Joseph J. Cech, Jr., University of California (Davis)
Richard D. Clime, Dodge Cove Marine Farm
Robert W. Corell, University of New Hampshire
Dudley D. Culley, Jr., Louisiana State University
Hussein M. El-Ibiary, University of Georgia
Charles E. Epifano, University of Delaware
Emanuel Epstein, University of California (Davis)
J. David Erickson, U.S. Trout Farmers Association
Arnold G. Eversole, Clemson University
Katherine C. Ewel, University of Florida
Stephen A. Flickinger, Colorado State University
John E. Foster, III, New Bern, North Carolina
Roger E. Garrett, University of California (Davis)
George G. Giddings, North Carolina State University
Donald S. Gilpatric, Acadia Aquacultural Enterprises, Inc.
Dana C. Goodrich, Jr., Cornell University
Malcolm S. Gordon, University of California (Los Angeles)
Wilbur Gould, Ohio State University
Donald G. Green, East-West Food Institute
Marilyn M. Harlin, University of Rhode Island
Gregory D. Hedden, University of Wisconsin
William H. Herke, Louisiana State University
Herbert Hidu, University of Maine (Orono)
Andrew G. Jordon, National Cotton Council of America
C. L. Kerns, University of Alaska
Niles R. Kevern, Michigan State University
Kurt F. Kline, California Polytechnic State University
George W. Klontz, University of Idaho
Richard L. Klosterman, Calaqua, Inc.
D. M. Knudson, Purdue University
Robert T. Lackey, U.S. Department of the Interior
Louis Leibovitz, Cornell University

E. W. Lewis, Marineland Farms, Inc.
George S. Lockwood, Monterey Abalone Farms
Richard A. Lutz, Yale University
Frank T. Manheim, U.S. Department of the Interior
G. C. Matthiessen, Marine Research, Inc.
Robert C. May, University of Hawaii (Manoa)
Ronald D. Mayo, Kramer, Chin & Mayo, Inc.
David C. McMillin, Olympia Oyster Growers Assoc.
James P. McVey, Trust Territory of the Pacific Islands
Paul Mulvihill, Aquaculture Research/Environmental Associates
Edward A. Myers, Abandoned Farm, Inc.
Scott H. Newton, University of Arkansas (Pine Bluff)
Robin M. Overstreet, Gulf Coast Research Laboratory
F. Ward Paine, SilverKing Oceanic Farms
John L. Plock, The Shelter Island Oyster Co., Inc.
Leo Polopolus, University of Florida
Fred Prochaska, University of Florida
Robert Putz, U.S. Fish and Wildlife Service
W. B. Quay, University of Wisconsin
John T. Quigley, University of Wisconsin
George Ray, Fish Breeders, California
Leo E. Ray, Catfish Farmers of America
Roland E. Reagan, Mississippi State University
Steve Rebach, University of Maryland (Eastern Shore)
William L. Rickards, University of North Carolina
Oswald A. Roels, University of Texas
Charles G. Scalet, South Dakota State University
R. S. Shallenberger, Cornell University
David I. Shapiro, TRACOR Sciences and Systems
Kenneth L. Simpson, University of Rhode Island
R. O. Sinnhuber, Oregon State University
Thomas R. Slough, ConAgra Fish Products
Roland W. Smith, Pond Construction Company, Vermont
Richard W. Soderberg, University of Wisconsin
R. N. Steele, Rock Point Oyster Co.
Robert R. Stickney, Texas A&M University
William Stone, Ohio State University
Robert C. Summerfelt, Iowa State University
Hugh A. Swingle, Conservation and Natural Resources,
   Alabama
George H. Taylor, Department of Marine Resources, Maine
O. W. Terry, State University of New York
Boyce Thorne-Miller, University of Rhode Island
M. R. Tripp, University of Delaware
Robert J. Valenti, Multi Aquaculture Systems, Inc.
J. Robert Waaland, University of Washington
J. R. Wall, George Mason University
Jaw-Kai Wang, University of Hawaii (Manoa)
B. K. Webb, Clemson University
Bruce T. Wilkins, Cornell University
Masashi Yamaguchi, James Cook University
Susan R. Zell, University of Southern California

CONTENTS

<u>PREFACE</u>

This study was undertaken at the request of the National
Oceanic and Atmospheric Administration and was supported by
that agency, the U.S. Department of Agriculture, and the
U.S. Fish and Wildlife Service. The study was conducted by
the Committee on Aquaculture, under the auspices of the
Board on Agriculture and Renewable Resources of the
Commission on Natural Resources of the National Research
Council.

The purpose of the study was to assess the state of
aquaculture in the United States so that recommendations
could be made to limit the constraints and enhance the
opportunities that relate to aquaculture. We hope these
recommendations will contribute positively to contemporary
efforts on the part of private enterprise, government, and
academia.

To study aquaculture, the Committee used an
interdisciplinary approach that included relevant aspects of
the natural sciences, technology, and the social sciences.
The Committee created study panels and conducted a workshop
on aquaculture. Information generated from these activities
was evaluated against written comments and suggestions
received from many individuals involved in various aspects
of aquaculture. Nearly 200 individuals representing a
cross-section of the U.S. aquaculture community were
consulted. The study also attempted to synthesize past work
in this field. Problems common to the different interests
involved in aquaculture were identified and they, in turn,
indicated areas where investment of resources and/or changes
in policy could be most effective.

We have tried to consider how we can best advise policy
makers, whether they control investment capital, government
activities, or the conduct of scientific research. The
Committee hopes that this report will also be a useful
compendium for all people concerned with aquaculture.

The Committee on Aquaculture is grateful to the many who
participated in this effort: the four supporting panels,
the workshop participants, and the many people who responded
to our request for comments. Special appreciation is

extended to Kent Price, of the University of Delaware for
his role as rapporteur, and to Bille Hougart, of the Board
on Agriculture and Renewable Resources of the National
Research Council, who was responsible for preparing this
report. The Chairman also appreciates the dedicated efforts
of the Committee secretary, Juanita Maclin, who cheerfully
kept pace with our demands, and of Barbara Davies, for her
able editorial assistance.

Don Walsh, <u>Chairman</u>
Committee on Aquaculture

<u>FINDINGS AND RECOMMENDATIONS</u>

Aquaculture--the human-controlled cultivation and harvest of both freshwater and marine aquatic species--has developed more slowly in the United States than other food sources, and is less advanced here than in other technologically advanced nations. Nonetheless, some species are being profitably cultured and efforts to expand the number of cultured species continue.

The political and economic priorities that should be given to aquaculture development are outside the expertise of this Committee. To place the topic in its proper perspective, however, it is important to note that current constraints on aquaculture development frequently discourage new initiatives, and that unless some of these constraints are overcome, progress will be further impeded.

Our analysis of the production, science, technology, economics, business, law, and administration of aquaculture, indicates that, in the United States, aquaculture will have only a minor impact on food production in the near term, in comparison with other food production systems. We have been unable to make a definitive statement about the exact future role of aquaculture as a source of food in the United States. It is, however, our considered opinion that in the long term, aquaculture will be a means of increasing protein supplies. We believe that aquaculture has the potential to contribute to increased food production. If this potential is to be tested, expenditures for current programs and for research and development must be increased.

Constraints on orderly development of aquaculture tend to be political and administrative, rather than scientific and technological. Advances are needed in all areas but for overall progress, the essential requirements are policy decisions and administrative actions. To claim that currently profitable aquaculture enterprises are critically dependent upon the development of an extensive national program would be exaggerated; nevertheless, development of aquaculture in general has been constrained by limited public support.

While it is unreasonable to expect that aquaculture should receive special exemptions from environmental and health regulations, it is not unreasonable to expect that aquaculture be given equity and parity with competing activities. The development of extensive aquaculture in coastal wetlands within the United States will be increasingly difficult owing to demands for alternative uses of those lands.

Although aquaculture has an active constituency, it has little political power within the framework of interest groups competing for governmental attention. Until recently there had not been a lead agency to direct, coordinate, and develop this field. To ensure a reasonable rate of development for aquaculture, a uniform set of aquaculture policies must be established. A lead agency must direct, guide, support, coordinate, and be responsible and accountable for activities among the relevant federal agencies. The recent designation of the U.S. Department of Agriculture (USDA) as the lead agency for aquaculture suggests that this need has been recognized.

PRODUCTION SYSTEMS

Production technology is currently adequate for profitable farming of catfish, trout, crayfish, baitfish, ornamental fish, and oysters. Similar levels of technology appear to be available for pen-reared salmon, and for the ocean ranching of salmon, but are not yet available for such popular species as marine shrimp or lobster or for marine finfish such as pompano. Specific problems affecting profitability still exist for some of these species. These problems are expected to be solved as the industry matures.

There are three primary constraints on the development of successful production systems that affect, to a greater or lesser degree, all types of commercial aquaculture. Two of these are related to gaps in the technology base and one is related to getting appropriate available technology into use in the industry. We recommend that priority consideration be given to:

*   <u>Improved technology for rearing cultured species during the early stages of their life cycles</u> (Chapters 1 and 2).

*   <u>Improved technology for formulating, processing, and delivering food and feed for cultured species</u> (Chapters 1, 2, and 3).

*   <u>Construction of demonstration facilities</u> (Chapter 1).

Knowledge pertaining to the biology of certain species, to culture technology, and to appropriate site selection is sufficient to permit some enterprises to be profitable. However, results from appropriate research programs in various scientific disciplines can help to reduce costs, open new opportunities for additional species, and provide greater assurance of economic profit and product quality.

Little information is available on the specific nutritional requirements of most culturable species at various stages in their life cycles although many cultured species are critically dependent upon the food provided to them. Appropriate diets can help to improve survival rates, growth rates, conversion efficiency, and disease resistance. We recommend:

- <u>Research to determine the nutritional requirements of culturable species throughout their life cycles</u> (Chapter 2).

Populations of aquatic organisms must be cultured in high density to be economically useful. Mass culturing frequently fails, however, because of insufficient knowledge concerning environmental conditions, behavior, and reproduction. To date, research has resulted in use of water temperature controls and gonadotropic hormones to stimulate reproduction in specific species. Further research is required. We recommend:

- <u>Research to improve understanding of mass culture in order to increase control over behavior and reproductive processes so that the necessary stocks of seed or juvenile organisms can be developed</u> (Chapters 1 and 2).

The great potential of improved breeding programs has yet to be realized for most culturable species of commercial importance. In the United States, genetic research has been primarily directed at the salmonids. Some crustaceans and molluscs are now amenable to the application of genetic techniques. Application of such techniques to aquatic organisms would improve product quality and the efficiency of the culture process. We recommend:

- <u>Research on the genetic characteristics of culturable species in order to develop strains possessing desired characteristics and to preserve natural and domesticated brood stocks</u> (Chapter 2).

Effective health management of cultured stock is
essential for successful culture. The control of disease in
culture systems by means of appropriate prophylaxis and
treatment is not yet adequately understood or applied.
Known pathogens can cause catastrophic losses, and further
culture experience will reveal additional threats from many
kinds of organisms. We recommend:

* <u>Research to improve knowledge of disease
organisms and to develop appropriate disease
control techniques</u> (Chapters 1 and 2).

The chemical and physical environment of cultured
species set absolute limits on the success of aquaculture.
Water entering aquaculture systems from ambient sources may
contain deleterious or toxic contaminants and destroy or
injure cultured populations. Optimum conditions of
temperature, salinity, sediment load, and water movement
exist for each developmental stage of each species.
Ignorance of these conditions has frequently precluded
success. Large competitive populations may retard growth,
waste food, consume oxygen, and produce wastes, thereby
reducing the efficiency of production. We recommend:

* <u>Research to determine the effects of water
quality and the physical environment on cultured
species so that techniques may be developed to
maintain an optimal environment</u> (Chapters 1 and 2).

The introduction of exotic stocks and species and the
frequent local transfer of stocks create a potential for the
introduction of disease, parasites, competitors and
injurious genetic strains. If a safe procedure for such
translocations can be developed, great improvements in
cultured products may result. We recommend:

* <u>Establishment of appropriate procedures for
registration of all exotic transfers for
aquaculture and for effective quarantine to prevent
introduction of disease</u> (Chapter 2).

ECONOMICS AND BUSINESS

The most serious economic constraints are: low prices
or limited markets for products; high prices or limited
availability of purchased inputs such as sites,
capital, unskilled labor, and trained managers and of
operating inputs, especially feed, water, seed, energy
and chemicals; and the great quantity of purchased
inputs necessary to produce output. The research we
propose here can enhance productivity and improve
understanding of markets.

Economic analysis of legal and institutional
effects on aquaculture development is needed as a guide
to public policy. Research should be focused on
species that have the greatest potential for acceptance
in U.S. or other markets. We recommend:

> • _Research to determine whether markets exist
> for species that are being considered for
> aquaculture. Such efforts should be concurrent
> with scientific research on those species_ (Chapter
> 3).

Because establishment of new ventures in aquaculture
will require significant market development that transcends
the ability of individual firms, we recommend:

> • _Federal support for public and private
> research on marketing aquacultural products_
> (Chapter 3).

Prices and availability of sites, labor, feed, seed, and
drugs have changed drastically in the past and will surely
change more in the future. In order to allow for adaptation
of technology over time, we recommend:

> • _Research on the relationship between potential
> aquacultural products and the land, labor, capital,
> and operating inputs required to produce the
> species. This research should also examine the
> relationship between scale of operation and input
> requirements_ (Chapter 3).

Aquaculture should be eligible for assistance from
programs of the Small Business Administration and the
Department of Agriculture. In particular, federal flood and
disaster insurance, crop insurance, and low cost financial
programs should be made available. Efforts should be
directed toward that segment of private aquaculture with
high potential profits and high accompanying risk.
Government programs generally available to support
activities in this category should be specifically available
to aquaculture enterprises. Aquaculture should be assured
the same tax, credit, insurance, depreciation, or
amortization benefits that are available to other forms of
small business. Therefore, we recommend:

> • _Financial programs that are now legally
> available to terrestrial agriculture be made
> equally available to those engaged in aquaculture_
> (Chapters 3 and 4).

LAW AND ADMINISTRATION

Aquaculture in the United States has lacked coherent
support and direction from the federal government.
Poor coordination, lack of leadership, and inadequate
financial support have traditionally characterized
programs relating to aquaculture.

On September 29, 1977, during the deliberations of this
Committee, the Food and Agriculture Act of 1977 (PL 95-113)
became law. The U.S. Department of Agriculture was
designated as the lead agency for aquaculture.

The designation of a lead agency was essential for
orderly growth of aquaculture within the United States.
Whether or not that agency is, or remains, the Department of
Agriculture, does not alter the thrust of our
recommendations. Our recommendations are intended to apply
to whatever federal unit is designated as the lead agency,
now or in the future.

In general, a lead agency should be responsible for
directing, planning, and coordinating U.S. aquaculture
efforts, and be accountable to the President and Congress on
the progress of aquacultural development. The lead agency
should seek funds from Congress in support of private and
public aquaculture activities and should represent the
interests of aquaculture before other federal agencies whose
regulations and policies directly or indirectly affect the
development of aquaculture. More specifically, we recommend
that:

- The lead agency take steps to develop
appropriate means to coordinate activities among
other federal agencies with aquaculture programs
(Chapter 4).

- The lead agency annually submit a report on
aquaculture in the United States to the President
and the Congress containing a review of existing
problems, progress to date, the allocation of
funds, and an outline for action (Chapter 4).

- The lead agency develop an integrated program
that recognizes the assets and skills of existing
aquaculture programs within other federal agencies.
Such a program should include a National
Aquaculture Plan that specifically considers the
resources of other federal agencies and contains a
timetable for action. (Chapter 4).

- The lead agency disseminate information
concerning federal and state permits and licenses

necessary for aquaculture operations and facilitate their acquisition (Chapter 4).

• The lead agency develop and promulgate guidelines for state use in enactment of laws and regulations relevant to aquaculture development (Chapter 4).

• The lead agency encourage the U.S. Environmental Protection Agency to review and establish specific effluent guidelines and standards for aquatic animal production (Chapters 2 and 4).

While the movement of new aquatic species into the United States and interstate needs to be controlled, many non-U.S. species have considerable potential as aquacultural candidates. Research and development efforts could be severely restricted without access to non-indigenous species. We recommend:

• The lead agency initiate with the U.S. Fish and Wildlife Service a review of the Lacey Act (18 USC 42) to: (1) identify species of importance to aquaculture; (2) establish a mechanism permitting entry of such species if the public interest will be best served by such entry; (3) develop and distribute to state authorities model regulations for interstate aquatic species disease certifications, and (4) compose and publish a listing of species whose entry into the United States is controlled by statute, but which are approved for use in aquaculture under specified conditions (Chapter 4).

Disputes may arise because ocean ranching involves public hatchery release programs for public benefit, as well as private ranching for profit. We recommend:

• The lead agency take legal and administrative steps to clarify the rights of both public and private hatcheries to harvest the ocean stocks they have planted (Chapter 4).

• The National Oceanic and Atmospheric Administration, through its office of Coastal Zone Management, issue guidelines to states for developing Coastal Zone Management plans that recognize the potential of aquaculture (Chapter 4).

• The Agricultural Extension Service of the U.S. Department of Agriculture be expanded to include services for aquaculturists (Chapter 4).

We recognize the potential strength of long-term
research in the Sea Grant program of the National
Oceanic and Atmospheric Administration and urge that
efforts be made to realize such potential.

The priority areas of research we have identified
are in nutrition, feed technology, genetics,
reproduction, health management, and various aspects of
aquaculture systems. Because problems in these areas
that relate to aquaculture require long-term research,
we recommend:

- _The aquaculture activities of the National
Oceanic and Atmospheric Administration's Sea Grant
program focus on the long-term research and
development requirements of aquaculture, rather
than on the short-term (three year), problem
solving program currently supported_ (Chapters 1, 2,
and 4).

CHAPTER I

PRODUCTION SYSTEMS

INTRODUCTION

In a world increasingly aware of the finite nature of its
natural resources, production of food is a pressing concern
for anyone dealing with the shape of the future. Aquaculture, the human-controlled cultivation and harvest of
both fresh and saltwater aquatic species, offers one means
for increasing that production. Aquaculture is based on an
assumption that proper management of controlled systems, by
permitting optimal use of input materials such as feed and
energy, can provide greater yield than is possible in
unmanaged natural systems. Invested materials, such as
feed, may return a substantial dividend on investment and
produce a more valuable market crop.

We are mindful that, although aquaculture can be
convincingly presented to be in the national interest,
aquaculture must compete with other food production and
recreational activities for federal support based on the
limited water, land, and money available. The relative
benefits contributed by aquaculture compared to the costs
invested is the final criterion on which U.S. aquaculture
must be judged.

Aquaculture in the United States began over one hundred
years ago, with salmon release programs designed to enhance
natural supplies. The present high level of enthusiasm for
fish farming is more recent, having begun several decades
ago. This interest was stimulated by a potential for
increased food, income, and recreation. Although economic
profit has been the strongest stimulus to date, the need for
food may, in the future, become the principal motivation for
development of aquaculture. In contrast with agriculture,
aquaculture is based on limited and relatively recent
efforts to apply science to increase production efficiency.

Total worldwide aquaculture production amounted to about
6 million metric tons (mmt) in 1975. Of this amount,
finfish represented about 66 percent, shellfish represented
about 16.5 percent, and aquatic plants about 17.5 percent.

9

Aquaculture production amounted to about 10 percent of total world fish production.

In some parts of the world, aquaculture production contributes significantly to human protein consumption. The greatest growth of aquaculture has occurred in countries with high levels of technology and risk capital, rather than in poor and protein-short nations. In Japan, production increased from 110,000 metric tons (mt) in 1971 to 500,000 mt in 1975; Israel obtains just under half of her finfish from culture, an amount that totalled 10,330 mt in 1974. These foreign achievements, and a series of laboratory-level and field successes in U.S. government and academic institutions, have increased enthusiasm for commercial aquaculture. Ocean ranching has become an important industry in Japan and in the U.S.S.R., having grown at an average annual rate of 7.3 percent in the last 20 years. These two countries combined released approximately 2 billion juvenile Pacific salmon from hatcheries in 1976. This represents a fourfold increase in the number of juvenile salmon released over the last 20 years and a potential harvest of about 40 million adult salmon by 1980.

Of the total world aquaculture production of 6 mmt in 1975, U.S. production amounted to only 65,000 mt. We estimate that world aquaculture production may reach 50 mmt by the year 2000. U.S. production may be over 250,000 mt by 1985 and, with proper support, could reach 1 mmt by the year 2000. U.S. production has not increased in recent years. This rate of development is slower than had been generally expected and has led to disappointment in aquaculture's contribution to U.S. food production.

Non-scientific factors are generally more responsible for this slow development than a lack of technical information. Nonetheless, research has concentrated on solutions to scientific problems, to the neglect of what now appear to be the most immediate problems--institutional constraints.

Although aquaculture for most species within the United States has not developed beyond the research stages, substantial progress has been made in several important sectors of commercial aquaculture, including the development of profitable production systems for trout, catfish, crayfish, baitfish, salmon, oysters, and ornamental fish such as goldfish.

Warm-water finfish culture is a significant use of the land and water resources in Alabama, Arkansas, Florida, Georgia, Louisiana, Mississippi, North Carolina, Oklahoma, South Carolina, Tennessee, and Texas. Isolated water resources such as geothermals in Idaho and Colorado permit culture of warm-water species in zones traditionally growing only cold-water fish. Products from these areas are sold to truckers who transport them live to distant markets, local fishing lakes, and processors. Fish produced with public money at public or private facilities are used for research, for supplementing natural populations, and for stocking public sport fishing waters. Channel catfish and bait minnows are the principal crops, but buffalofish, carp, tilapia, and various kinds of sunfish are also grown.

Coho and chinook salmon produced in publicly supported hatcheries on the Columbia River represent more than half of the salmon population in this large river caught commercially and for sport. Aquaculture, using hatchery techniques, has made progress toward growing pan-sized salmon in salt water. The ocean ranching of pacific salmon shows strong possibilities for success. Government hatching programs have been developed to mitigate losses of natural production from dam construction.

It is difficult to foretell to what extent our coastal areas will harbor aquaculture activities in the face of increasing competition from pollution, recreation and other pressures. Future growth and development should include both intensive aquaculture, in which the organism lives in a highly controlled environment for all or part of its life cycle, and extensive aquaculture, in which the organism lives in a somewhat controlled environment for all or part of its life cycle. We note with approval efforts to master the intensive culture of certain aquatic species. These culture systems offer great potential for using various thermal energy sources that might otherwise go to waste.

Although we cannot estimate the best future balance between intensive and extensive aquaculture production systems, we are confident that aquaculture, as a whole, can make a large contribution in future years. It is in increasing gross food production and in the development of alternative uses of renewable resources that aquaculture can provide the greatest benefits. Better uses of water supplies and of heretofore wasted thermal energy supplies are possible. While aquatic food production systems can be pollutant sources, the waste can be beneficial if properly handled. Aquaculture systems offer considerable promise as a sewage treatment tool.

In this chapter, we first examine aquaculture production systems that are currently used in the United States and then briefly examine some additional systems that might be profitable in the future. By examining the production systems for various species or groups of species, common constraints affecting aquaculture can be identified. These constraints are then discussed in the later chapters.

Problems identified in the species studied relate to
scientific knowledge and its application, economics, and
government regulations. These problems are interrelated and
exist within a larger framework of public policy management.
The management problems do not emerge from the individual
studies but will be addressed in the final chapter.

Although this report focuses primarily on food fish,
ornamental and bait fish have been included because they are
economically significant products of the aquaculture
industry. Conversely, due to time constraints, numerous
food species that are not economically significant have been
omitted.

The descriptions of the production systems are not
intended to be exhaustive but to provide general information
on the state of aquaculture technology. A bibliography is
included at the end of the report for readers desiring more
detailed information on culture of various species.

CURRENT PRODUCTION SYSTEMS

Salmon and Trout

All trout available to the U.S. market come from farming
operations. Over 100 farms produce trout as their primary
crop, and over 700 small enterprises grow and sell trout to
stock private waters. Large scale commercial operations are
of recent origin: in 1954 only 450 mt of these fish were
produced; in 1972 production amounted to 13,400 mt.

Most commercial trout farms are in Idaho and Montana,
with smaller numbers in Washington, Oregon, and California.
Some production also occurs in Minnesota, Wisconsin,
Michigan, Ohio, and in a few southeastern states.

Salmon aquaculture began in California in 1872, but
contributed little to increasing natural runs of salmon
until 1960, when research began to provide new information,
particularly about nutrition and disease control. Since
then, significant catches have resulted from fish that were
raised in government hatcheries on the Columbia River and
were later released to spend most of their lives in the
ocean. These hatchery fish account for more than half the
coho and chinook salmon fish caught in that large river
system.

Salmon and trout aquaculture can be separated into two
categories: (1) feedlot rearing, where fish are raised to
market size in captivity; and (2) ocean ranching, where fish
are released into the ocean as juveniles and harvested as
maturing adults. Feedlot rearing is an established U.S.

industry in freshwater, where primarily trout are grown and a developing industry in salt water, where primarily Pacific salmon are grown. Ocean ranching is conducted by state and federal agencies to support fisheries and a developing aquaculture industry (see Potential Production Systems).

Salmon and trout require high-quality water and high protein food. Scarcity of processed foods affects feedlot rearing more than ocean ranching because ranched fish are released into the ocean at a small size to graze on natural food.

Ten or more species of salmon and trout are currently considered suitable for aquaculture. However, modifications in technology are necessary to satisfy the ecological and behavioral requirements of individual species. Modifications are also needed to meet production requirements for feedlot rearing and ocean ranching. A detailed assessment of production systems will not be attempted here. Instead, the production cycle will be discussed in general terms to identify important problems.

Depending on species and environment, salmon and trout mature between two and six years of age. The number of eggs is relatively low in comparison with other species of food fish. Artificial spawning is easily accomplished and results in a high rate of egg fertilization.

Eggs are incubated efficiently but must be chemically treated for protection against infections. Certain chemicals commonly used for rearing eggs are not approved by regulatory agencies. Approval of existing prophylactic methods or development of alternative methods is urgently needed.

Conventional methods for raising larval salmon and trout often produce undersized juveniles that may have a reduced capacity to survive, particularly if released into the ocean. New incubation systems to provide improved conditions for larvae are generally more costly than conventional systems, and an improved technology for rearing larval stages is needed.

A variety of tanks, raceways, and ponds are successfully used to hold juveniles. Particular configurations can be used to resolve problems related to water supply and hatchery site development or to adapt a rearing system to operations that may involve specific technologies of water reuse, feeding, fish handling, waste disposal, and disease control. The demand for improved technology on water treatment for control of pathogens, metabolites, dissolved nutrients, temperature, suspended solids, and organic wastes will increase as fresh water becomes more scarce.

Greater efficiency is needed in feeding, grading,
treating for disease, and transporting large numbers of live
fish. Artificial feed is expensive, often difficult to
transport and store, and does not produce fish whose color,
texture, and flavor are equivalent to that produced by
natural food. Genetic selection has often resulted in
inbreeding and loss of adaptive genetic variability.

Development of salmon brood stocks is especially
difficult where natural stocks have been decimated from
over-fishing or where deterioration of fresh-water habitat
has occurred. Scarcity of brood stock, though not a serious
problem with trout, is a serious impediment to growth of
salmonid aquaculture.

Warm-water Finfish

Channel Catfish

The channel catfish is the principal farm-raised species in
the United States. Since 1955, when technology for reliable
spawning and fingerling production was developed, the
industry has grown rapidly. By 1970, large scale production
was commercially viable. The area devoted to catfish
increased from less than 50 hectares (ha) in 1955 to 18,809
ha in 1975. Production was 4,000 mt in 1969 and 38,000 mt
in 1975. Catfish are raised by approximately 2,000 farmers
in 13 southern states and are processed by 12 different
firms. Three states, Mississippi, Arkansas, and Louisiana,
produce 80 percent of the total.

Increased costs of production, related to labor, feed,
equipment, and water pumping, have resulted in a decrease in
the area devoted to catfish culture. Nonetheless, in the
past five years, average yields have risen from 1,344 to
2,017 kilogram per hectacre (kg/ha). At a farm price of
$1.10/kg, the 1975 crop was worth $41,731,800.

Spawning of channel catfish is usually accomplished by
stocking freshwater ponds with equal numbers of males and
females. Spawning receptacles, such as milk cans or kegs,
are provided. Channel catfish begin spawning when water
temperature is above 21°C. In some hatcheries, the egg mass
is removed to a hatchery trough provided with a paddle wheel
for agitation and aeration to simulate the fanning action of
the male catfish in nature. However, in most cases, the
eggs remain in the container in the pond and the male
attends to their hatching.

Requirements for incubation and hatching are not
particularly stringent and successful operations have been
carried out in many geographical areas. Existing technology

14

related to spawning and incubation is suitable for the immediate needs of the industry.

Recently hatched fry are counted and moved to a rearing trough in the hatchery or to a pond. Fry are fed pelleted diets at two- to four-hour intervals. Fry may be grown to fingerling size in about three months or may be moved at any time to ponds.

Rearing ponds for fingerlings vary from 0.04 to 2.0 ha. The pond is filled with water immediately before being stocked with fry in order to reduce the number of predatory insects present. Ponds may subsequently be treated with motor oil and kerosene twice per week to kill air breathing insects. Fry are stocked at 50,000-375,000/ha, and fed 4-5 percent of body weight daily with a commercial pelleted feed.

Major problems in rearing fry to fingerlings are caused by disease agents for which no immunological procedures have been developed. Few chemotherapeutic materials are effective or cleared for use on food fish by federal regulatory agencies.

Fingerlings (15-cm) in growout ponds (stocked at 7,500/ha) can be grown to 0.45 kg in 180 days from April to October. A commercial pelleted ration with 35 percent protein is used; the amount fed daily is gradually reduced during the growout period. Best growth occurs when temperatures range from 22°C to 30°C, although the fish will feed and grow slowly even at 10°C. Floating feed allows farmers to see the response of the fish to their food and thereby allows them to monitor the well-being of the fish.

The largest single problem for catfish aquaculture derives from oxygen depletion. Such depletion places stress on fish which predisposes them to disease.

Special systems for growing fingerlings to market size include cages and raceways. Cages are floated in ponds, reservoirs, or slow-moving streams. The stocking rate is 300-500 fish per cubic meter cage, but, in standing water bodies such as ponds, the volume of water determines the total fish weight that can be grown, and caging does not affect this relationship. Caged fish that are fed a nutritionally balanced pellet grow to market size in essentially the same time as do fish in ponds.

Raceway culture of catfish has been adapted from that of rainbow trout and may be integrated with winter culture of trout, allowing double-cropping in the same facility. Raceway units are usually located on slopes, and reservoirs are at the upper and lower ends of the raceway. Water is

recirculated through the two ponds and the raceways.
Catfish fingerlings are stocked at about 2,000 per unit and
grown to market size of .45 kg in six months.

Harvesting from ponds is done by hook-and-line, seining,
draining, or by a combination of these methods.
Considerable progress has been made to mechanize seining and
handling of live catfish. Market-size fish are seined
(using selective meshes), and the catch is loaded onto
hauling vehicles by booms containing weighing apparatus. At
the processing plant, the fish are killed by electricity,
and various combinations of machines and hand labor are used
to prepare the marketed product.

More efficient mechanized methods are needed to
concentrate and harvest catfish in ponds built in rolling
terrain where obstructions to seining may exist.

The profitability of catfish farming declined in the
early 1970s as production costs (primarily feed) increased
and production levels remained low. The industry was also
plagued with marketing problems due, in part, to lack of
consumer interest in the product. Profitability has
rebounded in the past three or four years. Average
production per unit area has increased steadily, and prices
paid to farmers have increased dramatically. Further
increases in productivity can be expected in this industry
as new technologies are incorporated.

Bait Minnows

Bait minnow culture began in the United States in the early
1900s and attained a high level of development. This form
of aquaculture ranks close to channel catfish in area
devoted to production in the southern United States; 9,710
ha in 1968 and 16,290 ha in 1975. Minnows produced in the
South are marketed throughout the United States. The
wholesale value of bait minnows produced in 1975 ($4.40/kg)
was $43,005,600.

The golden shiner (_Notemigonus crysoleucas_) is the
primary bait fish that is cultured, although others, such as
the fathead minnow (_Pimephales promelas_) and goldfish
(_Carassius auratus_), are also grown.

Spawning and reproduction of the golden shiner occurs in
freshwater ponds. Ponds are up to 32 ha in area and usually
have rows of vegetation (such as ryegrass) along the
shoreline upon which fish deposit eggs. One-year-old brood
fish are selected to minimize incidence of an ovarian
parasite, _Plistophora ovariae_. This parasite is a major
problem in older female minnows.

Eggs are unattended by brood fish and spawning mats
containing eggs are usually placed in nursery ponds where
hatching occurs within four to seven days.

Young shiners feed on plankton. Plankton blooms are
stimulated by adding fertilizer. Finely ground supplemental
feed also may be fed to very young fry.

When shiners reach 20-25 mm, they can be seined and
moved to rearing ponds in densities of 375,000-625,000/ha.
As fish grow, feed is gradually increased to a maxiumum of
20 kg/ha/day. Yields of marketable shiners reach 600
kg/ha/year. Minnows that are too small to sell at harvest
during December and January are restocked at 125,000/ha,
grown for about two months, and then sold before spawning
time.

Technology for culturing shiners is essentially
adequate. However, formulations of appropriate feed for
bait fish must be developed.

Shiners are harvested by seine, and care must be taken
to avoid injury to these delicate fish. They are generally
handled in cool weather to avoid stress. Oxygen, ice, salt,
and bacteriostats are used to facilitate transportation.

Systems are adequate for handling bait minnows, but a
variety of chemotherapeutics must remain available for use
by the fish culturist. Several useful chemicals are
currently under study for re-registration by regulatory
agencies. If these chemicals are banned for use on food
fish, their use on the bait minnows might also be
prohibited.

Buffalofish

About 454 mt of the largemouth buffalofish (<u>Ictiobus
cyprinellus</u>) are produced annually on about 1,000 ha, mainly
in Arkansas. This food fish, produced extensively in
irrigation reservoirs, yields only about 112 kg/ha.
Cultured intensively with fertilizers and feed, buffalofish
can yield 900 kg/ha. Retail price is about $1.10/kg, and
the gross value of the 1975 production was about $500,000.

At present, monoculture of buffalofish is not a
profitable use of aquacultural land. Some farmers are using
buffalofish in polyculture with channel catfish, but recent
discoveries indicate that, in ponds with low inherent
fertility, buffalofish compete with channel catfish for
supplemental feed.

Technology for spawning and rearing buffalofish is well-established. Refinement of polyculture methods is required, and product development is needed to more profitably market the excellent flesh.

Carp

The common carp (<u>Cyprinus carpio</u>), is cultured noncommercially in irrigation reservoirs and to a small extent for sale as biological weed-control agents. In 1972, total production was 118,000 kg. The common carp, grown in the United States since the 1870s, has never gained broad acceptance as food and producers can only obtain prices of about $.57/kg.

Since 1963, three species of Chinese carps have been introduced for testing as farm fish. These are the grass carp (<u>Ctenopharyngodon idella</u>), the silver carp (<u>Hypophthalmichthys molitrix</u>), and the bighead carp (<u>Aristichthys nobilis</u>). Of these, the grass carp is best known by farmers for its effective control of aquatic macrophytes in fish ponds.

The silver carp has been grown only under experimental conditions in the United States, but has shown great potential for controlling excessive phytoplankton in catfish culture ponds. As many as 3,470 silver carp have been added per hectare (to ponds already stocked with channel catfish) and the silver carp have not competed with catfish for food and have yielded 1000 kg/ha. This represents a significant increase in production efficiency and a considerable addition to the food supply. Silver carp have been sold in the retail market at $2.80/kg. Given their potential for improving water quality in traditional catfish ponds, silver carp should become an important aquaculture species in the future.

The bighead carp, a zooplankton feeder, competes with catfish for supplemental feed and is less useful than silver carp in polyculture. In one catfish pond, bighead carp at 125/ha produced 400 kg/ha over an 18-month period with no evident effect on catfish production.

Development of carp culture in the United States could be greatly expanded through aggressive marketing to overcome the "trash fish" image usually associated with the common carp. Chinese carp should be differentiated from the common carp.

Tilapia

Blue tilapia (<u>Sarotherodon aurea</u>), an Israeli species
introduced to the United States about 1957, has been
successfully grown in polyculture with catfish. This fish
feeds on algae, detritus, and waste feed and produces up to
1500 kg/ha without reducing catfish production. Fish of
marketable size command retail prices as high as $3.50/kg.
Currently, large numbers of fish do not reach marketable
size because reproduction occurs in the culture pond.
Recent experimental programs in monosex-culture have shown
the feasibility of producing crops in which virtually all
the fish are marketable. Monosex populations of tilapia can
be produced by hybridization, by sex reversal using oral
administration of hormones, and by manual separation of the
sexes.

Tilapia are produced in Mississippi, Alabama, Florida,
California, Utah, and Colorado. The main constraint to
production is the fish's inability to tolerate water
temperatures much below 8°C. Tilapia must spend the winter
in water warmer than normal winter temperature or even grown
in heated water. Other requirements are economical
production systems for monosex populations and more precise
data on nutrition of tilapia.

Ornamental Fish

The major producing centers for ornamental fish are in West
Germany, England, Holland, Denmark, Belgium, Japan, Hong
Kong, Singapore, and the United States. The United States
represents 50 percent of the world market, with some $700
million spent annually on aquarium fish and supplies.
Florida, the world's largest single breeding center for
ornamental fish, currently has some 150 tropical fish farms.
In 1972, these farms produced 97 million fish. In that
year, Florida imported an additional 53 million ornamental
fish, so the combined weight of local and imported fish was
approximately 10,200 mt. The estimated retail value of the
fish was $300 million.

Molluscs

The bivalve molluscan industry is almost entirely dependent
upon natural catch. As a production system, the industry is
troubled by decreasing yield. Overfishing, disease, natural
disaster, predation, pollution, and loss of natural grounds
have combined to almost destroy this valuable food source.
Considerable effort is being expended in hatchery operations
to revitalize the industry, but massive planting programs
alone are not apt to offset the many problems that confront

the industry. With diminishing natural populations, further
development of high-performance harvest and processing
equipment is of questionable value.

Oysters

Oyster culture has been practiced in some parts of the world
for thousands of years, but began in the United States only
in the late 1800s. The levels of complexity of oyster
culture vary; some systems are based on minimal human
intervention, others on intensive husbandry.

Oyster production in the United States has declined
greatly from the record 69,000 mt in 1908; production was
34,000 mt in the late 1940s and 22,000 mt in 1973. More
than 40 percent of 1973 production came from farming
operations most of which are very simple and can barely be
called culture. Farm operators collect spat on some
suitable substrate. The substrate with the young oysters is
then deposited on bottoms where the spat grow to harvestable
size. The oysters may be thinned or redistributed as they
grow. Yields of untended oyster beds could be 7 to 20 times
greater if more advanced culture were used. The future of
this industry depends on expansion of culture systems. With
adequate markets, acceptable costs, and revisions in many
state laws to favor private culture, production could
quadruple in the next ten years.

Private aquaculture predominates in some areas such as
Long Island Sound and the Pacific Coast states. Public
fisheries are active in the Gulf and East Coast states.

Although four species of oysters are currently cultured
in the United States, the 1975 oyster crop, which amounted
to 24,000 mt, was comprised almost entirely of two species,
the American oyster (_Crassostrea virginica_) and the Pacific
oyster (_C. gigas_). Very small quantities of the European
oyster (_Ostrea edulis_) and the Olympic oyster (_O. lurida_)
are grown.

Oyster culture technology has advanced substantially in
recent years. Techniques have been developed for raising
larvae in hatcheries, and all stages of the oyster's life
cycle can now be controlled. Artificial spawning techniques
permit seed to be produced as needed, all year, thereby
eliminating the uncertainty of natural reproduction.
Unfortunately, the cost of growing seed to marketable size
in the hatchery currently prohibits commercial production.

Oysters normally develop gametes in spring and spawn
during summer. Spawning is related to temperature. The
spawning system is relatively simple. When water

temperatures are raised in the hatchery, gametes develop.
Ripe oysters can either be allowed to spawn or can be placed
in cooler water to delay spawning.

Larvae are reared in specially designed tanks in
hatcheries. Under experimental conditions, the right
combination of algae diets has shortened the larval stage to
nine days.

A large cost in oyster hatcheries is growing algae to
feed the larvae. Studies are underway to look for
artificial diets which could possibly eliminate the cost and
space necessary to culture algae.

Newly set oysters are extremely vulnerable. They are
easy prey to many predators. If seed oysters are not placed
on a suitable bottom they can easily be smothered by silt.
Oysters can also be seriously affected by freshwater
runoffs.

Once oysters are 50 mm or greater, losses from predation
are less severe, but at this stage losses can be caused by
diseases. Systems to protect oysters from diseases are
needed. Advances have been made in developing genetic
strains resistant to diseases, but full-fledged field
testing has not been done.

Almost all oysters are grown to market size on a
substrate or bottom. The time required for oysters to reach
market size (75 mm or greater) depends on water temperature.
In New England, for example, four to five years are
required, but, in the Gulf of Mexico, less than three years
are needed.

Off-bottom culture produces marketable oysters in about
one-half the time of bottom culture. Production systems to
economically grow oysters off-bottom should be developed.

Three promising new methods for growing oysters to
market size are currently being studied. These are closed
systems, culture systems using recycled wastes, and systems
using artificial upwellings. The economic feasibility of
these systems has not been determined.

Oysters are harvested by a variety of methods controlled
by state law. These methods include power dredging in Long
Island Sound, by hand tongs, patent tongs, and sail dredging
in the Maryland portions of the Chesapeake Bay, and by large
escalator dredges on the West Coast.

The majority of oysters are shucked by hand for the
fresh market. There is a definite need for the development

of a machine capable of shucking raw oysters. Modern
methods are available for cooking oysters for market.

Other Bivalves

In the United States, oysters are the only bivalves that are
profitably cultured. Research and a limited amount of pilot
testing has been conducted for clams, scallops, and mussels.

   Although hard clams (Mercenaria mercenaria) have been
reared in hatcheries, when transplanted in the wild, they
have rarely reached market size. The Manila clam (Venerupis
japonica) has also been reared in the hatchery and grown to
market size on plankton produced in waters from artificial
upwelling. Attempts to seed intertidal areas with Manila
clams in Washington have had limited success. Surf clams
(Spisula solidissima) have also been reared in hatcheries.
In all cases, no increases in the commercial catch have been
traced to aquaculture ventures related to clams.

   Limited supplies and growing demand provide incentives
for raising clams. Techniques developed from oyster culture
are available for clams, thereby making farming at least
technically feasible. Clam seed has been produced in
hatcheries since about 1970. Commercial firms have produced
small quantities of hard juvenile clams for planting. A
North Carolina hatchery has the capacity to produce over
four million seed clams a year for planting. Lack of
suitable grounds in which to transplant seed is a principal
constraint.

   The small demand for mussels (Mytilus edulis) in the
United States has provided little incentive for mussel
farming. Slowly increasing demand may accelerate
development of a culture system. Mussels are a major
aquaculture industry in some countries, such as Spain and
the Netherlands. However, the labor-intensive methods used
in those countries are not easily adaptable to the United
States, and modifications will have to be developed.

   Three scallops (the bay [Argopecten irradians], calico
[A. gibbus], and sea [Patinopecten yessoensis]) have been
reared to metamorphosis in hatcheries. Attempts have only
been made to rear the bay scallop to market size. Cage
culture can produce scallops in six months, making the
scallop a prime candidate for aquaculture. Scallop supplies
are erratic and usually insufficient. Relatively little
work has been done with these species, although it is known
that spawning can be induced and larvae of some scallops can
be raised to adult size. Pilot testing of laboratory
procedures will be required.

Abalone

The abalone (<u>Haliotis</u> <u>rufescens</u>) is the only gastropod
cultured in the United States. The Japanese have cultured
abalone for many years, mainly to re-seed natural stocks.
Currently, several commercial abalone farms, located in
California, have successfully produced seed abalone.
Ability to grow seed to market size has not yet been
commercially demonstrated.

Young abalone graze on seaweeds, and systems that
provide artificial feed have yet to be tested. Some grow-
out systems have been designed but further testing and
development is necessary.

Unless some protection from predators can be provided,
the potential for farming abalone in many coastal areas is
questionable.

Crustacea

Among crustaceans raised in the United States (crayfish,
lobster, freshwater shrimp and marine shrimp), only crayfish
are raised profitably. The high demand and prices for
crustaceans have stimulated substantial investments in
research and development from the private sector.

The popular conception of farming the seas or converting
natural waters into giant sea farms does not appear to be
feasible for crustacean farming. Production of crustaceans
from uncontrolled natural systems is low and the demand for
these natural waters for recreation and commercial fishing
is great. The ability to control predation, water quality,
and competition is a necessary part of crustacean farming.
This control can probably be realized only in ponds or
tanks.

Crayfish

The commercially important species are the red swamp
crayfish (<u>Procambarus</u> <u>clarkii</u>) which represents 90 percent
of the total catch, and the white river crayfish
(<u>Procambarus</u> <u>acutus</u> <u>acutus</u>) which represents the remaining
10 percent.

Lousiana produces 99 percent of the crayfish in the
United States and consumes 85 percent locally. The industry
includes both wild-caught and pond-raised crayfish. All
crayfish farming is done by local landowners; large
corporations have not entered the field. The bulk of
Louisiana's wild-caught crayfish come from the Atchafalaya

Basin; the Basin's harvest has fluctuated from 90,800 kg in 1959 to 4,500,000 kg in 1965, depending on water levels and temperatures. Production is uncertain and unmanageable with bumper crops being produced in the Basin about two years out of five.

Crayfish are sold live or as peeled tail meat. Approximately 40 processing plants operate in south Louisiana. Crayfish farming is an integral part of Louisiana's economy and in 1977, some 6,806 mt were produced from about 20,235 ha. Some additional 81,000 ha have been identified as potential sites for crayfish farming in Louisiana. Based on increases in past years, the hectarage devoted to crayfish farming could double by 1985.

Crayfish farming is conducted in three types of ponds: rice-field, wooded, and open. In rice-field ponds, crayfish are rotated with rice: the rice is harvested, the ponds are reflooded; crayfish feed on decaying rice stubble. Ponds in wooded areas are poor for crayfish because of poor wind circulation, resulting in low dissolved oxygen. Open ponds are constructed solely for crayfish farming.

The procedure for farming all three types of ponds is generally the same. Crayfish are stocked in ponds in late May or June. Brood stock, usually bought from a dealer, is stocked at rates of 25 to 50 kg/ha, depending on the amount of vegetation and the number of native crayfish present. Once stocked, the crayfish burrow, or go underground. After the crayfish has burrowed, the water is drained from the pond to control predators and to encourage the growth of aquatic vegetation upon which crayfish feed.

Female brook stock produce young while in the burrow, and human intervention is unnecessary. Each female lays between 100 and 700 eggs, with 200 to 300 as an average. Eggs are attached to the abdomen of the female where they are incubated until they hatch. Larval crayfish are, in fact, miniature crayfish, and do not undergo morphological changes into various larval stages as do <u>Macrobrachium</u> and Penaeid shrimp.

When young crayfish are found in burrows (usually in September or October) the ponds are flooded to release the young. Young crayfish forage on native plants such as alligatorweed, water primrose, and smartweed. Under favorable conditions--adequate food, high-quality water with dissolved oxygen at or above 3 ppm, and warm water with temperatures above 20°C--newly hatched young may reach maturity in three months.

Crayfish are harvested by trapping. Approximately 25 funnel traps of chicken wire are used per ha. Shad or fish

heads are used as bait. Trapping may begin in late November
and continue through May of the following year. Ponds with
good production might yield from 225 to 900 kg/ha, and
yields of over 1,720 kg/ha have been reported.

Crayfish farming is limited by three major constraints:
(1) harvesting techniques are labor intensive and may cost
up to one-half the total income; (2) production per ha is
low compared with production of catfish; and (3) labor to
peel crayfish tails is costly, but a peeling machine which
shows promise has been introduced.

Advantages of crayfish farming are: (1) required
capital investment is low as compared with catfish farming;
(2) no special feeds are required; (3) little management is
needed; and (4) brood crayfish need be stocked only once.

Marine Shrimp

In the 1950s, intensive culture of marine shrimp (Penaeus)
was economically profitable in Japan, and interest began to
develop in the United States. Laboratory techniques, based
on the Japanese model, were developed for producing large
numbers of seed animals for planting in ponds or other
enclosures. Methods were also developed for holding shrimp
until they grow to marketable size. These experiments were
encouraging, and a number of commercial firms launched pilot
enterprises. In some years production has reached nearly
500,000 kg.

However, laboratory and pilot-scale procedures have yet
to be commercially viable on a large scale, and marine
shrimp farming is currently not a profitable industry in
this country. One by one, companies have either
discontinued their development activities or have moved
their operations to Latin America. Factors contributing to
the failure of marine shrimp aquaculture in the United
States are: inadequate knowledge of nutritional
requirements, inadequate reproductive control, unsuitable
climate for year-round production, high costs of food and
labor, and legal and social constraints as they relate to
water access.

Cultured marine shrimp generally have not been matured
and spawned routinely in captivity. Wild gravid females
used as spawners must be collected and taken to hatcheries
where spawning is induced. In some cases, immature wild
males and females of spawning age are collected and held in
systems where maturation, mating, and spawning occur.
However, both the number of eggs and the survival rate of
larvae are negatively affected by this procedure.

Collection of wild spawners requires a large amount of boat-time and is, therefore, costly. Improvements in collection procedures have already been implemented and further advances are dependent on additional knowledge from research on reproduction. Dependence on wild spawners prohibits the type of genetic improvements that can be obtained with species that can be domesticated.

Egg incubation and hatching procedures are routine for marine shrimp when high quality seawater is available and good hatchery management procedures are used. Rearing procedures are well established for marine shrimp larvae and involve careful control of temperature, salinity, oxygen, and other water quality conditions. Algae and brine shrimp are required as food.

Larval-rearing procedures are expensive and contribute significantly to the cost of raising shrimp. Older, less expensive rearing methods can be used but these require greater initial capital investment and are less dependable. Major costs are associated with both the algae and brine shrimp and recent demand for brine shrimp has exceeded the supply. The development of suitable artificial feeds to replace algae and brine shrimp and the development of efficient systems for the mass production of brine shrimp would improve the efficiency of larval culture.

Two distinct systems are used to rear marine shrimp after they have been removed from the larval rearing tanks. These are semi-natural rearing ponds and raceways. Raceways are used to increase the survival rate and to control temperature, thereby extending the length of the growing season. The growing season is about seven months in natural ponds in the southern United States. Refinements of both systems are needed to reduce costs. Either system can be used to rear juvenile shrimp at high densities, thereby reducing space requirements.

Ponds are stocked with one or two shrimp per square foot and the shrimp are fed formulated dry rations at set rates. Typically, some water is exchanged during the growout period and predator control is exercised by filtration of incoming water. The universal problem with marine shrimp farming is that either a large poundage of small shrimp or a small poundage of large shrimp can be produced, but not a large poundage of large shrimp. Occasional production of up to 4,500 kg/ha per crop of large shrimp has been reported.

Few studies have been conducted to identify the optimum ecological conditions for marine shrimp. Most public research facilities do not have adequate pond space for such studies and commercial firms have apparently avoided such work because of the practical difficulties in controlling

pond ecology. However, major increases in production might
be realized through the refinement of methods for
environmental control.

     Harvesting is accomplished by draining and collecting
the shrimp in nets as they leave the pond outlet. This
procedure is well established, relatively efficient and
requires little improvement. Processing methods used by the
shrimp industry are generally suitable.

Freshwater Shrimp

Freshwater shrimp (Macrobrachium) culture is approaching
profitably in Hawaii. Efforts to culture these shrimp are
also underway in Puerto Rico. Several industrial groups are
conducting pilot studies to determine the economic
feasibility of farming these shrimp in the continental
United States. However, markets are still small and
production is modest. Rising production costs and the
scarcity of appropriate aquaculture sites may greatly slow
development of this species.

     Research on culture of freshwater shrimp was originally
conducted by scientists of the Food and Agriculture
Organization (FAO) of the United Nations in Malaysia, in
1959. Since then, research conducted in Hawaii has resulted
in a system for raising laboratory produced larvae to
commercial size in a seven month period. Production of
about 1.4 mt/ha/year has been achieved in Hawaii. Still,
less than 100 ha worldwide were devoted to the culture of
these animals in 1977. Indications are that several hundred
hectares may be in production in Hawaii within the next
several years.

     Freshwater shrimp mature, mate, and spawn naturally in
captivity without special handling. Maturing males and
females are placed in tanks where mating and spawning occur.
Females incubate the eggs, which are attached to their
bodies after spawning. When eggs hatch, the larvae are free
swimming and can be collected from spawning tanks.
Successful hatching depends on control of the quality,
temperature, and salinity of the water.

     Freshwater shrimp are reared easily at low densities.
Larvae and young require brackish water, but juveniles are
grown out in fresh water. Larvae are reared in tanks
designed to keep them suspended in the water column and
brine shrimp and algae are provided as food. At present
brine shrimp are essential for larval culture. The high
costs associated with algae and brine shrimp as food for
marine shrimp are also true for Macrobrachium culture.

Disease problems are frequently encountered during
larval stages. Research that resulted in increased ability
to rear larvae in high-density conditions and development of
acceptable artificial feeds for larvae would decrease the
costs of larval rearing.

Pond temperatures suitable for growth are found only in
the warmest parts of the United States. The development of
temperature control systems such as solar heating would be
useful in extending the length of the growing season. The
use of water from thermal discharges should also be
explored.

Juvenile and adult shrimp are stocked together in ponds
and stock is replenished at intervals of two to four weeks.
Chicken feed or similar organic feed-fertilizers are
provided. Pond management is difficult because of the
aggressive behavior of freshwater shrimp. A small
percentage of the population dominates the other shrimp in
the pond and grows more rapidly. If the dominant group is
removed, other shrimp become dominant. Hence, shrimp of
marketable size must be removed from the pond at regular
intervals.

The largest shrimp are removed for marketing and the
remainder are returned to the ponds. Smaller shrimp are
frequently injured or killed during the process.
Development of better techniques for harvesting large shrimp
would be of great benefit. Industry is experimenting with
methods of herding (attracting or repelling) and with
techniques for sorting live shrimp.

Shrimp are sold either fresh or frozen, and processing
does not pose any problems.

Lobster

Supplies of the northern lobster (_Homarus_) are inadequate to
meet demand and therefore the species commands high prices
and is attractive to aquaculturists. Lobster culture began
in the 1800s in the United States. For many years, larvae
and juveniles were released into the ocean to augment
natural production. This procedure was not demonstrated to
be effective and lobsters are currently raised in captivity
and grown to market size in two to three years. Growth is
accelerated through elevated water temperature and supply of
ample artificial feed. However, due to the high cost of
food and the cost of separating lobsters (which are
cannibalistic), culture procedures are not yet profitable.

Costs may be reduced if research on lobster culture can
provide methods for further accelerating growth and can

provide solutions to other problems associated with this
species. However, profitable commercial culture is probably
some years away.

Maturation, mating, and spawning of lobsters occurs in
captivity without extensive manipulation of environmental
factors. Lobsters must be several years of age before they
spawn and the number of eggs produced by captive lobsters is
usually less than the number produced by wild lobsters.

Eggs are attached to the abdomen of the female where
they are incubated for 4 to 8 months depending upon water
temperature. No serious problems occur during the
incubation process. A variety of larval rearing procedures
have been used successfully.

Procedures for rearing juveniles and adults are
essentially similar. The aggressive behavior of lobsters
dictates that they be reared in individual compartments to
avoid cannibalism. Despite considerable effort on the part
of engineers, culture systems with individual compartments
for lobsters greatly increase the costs of supplying feed
and oxygen and of removing wastes.

Juvenile and adult lobsters are generally fed live or
fresh clams or fish. Experiments with less expensive
compound diets are underway but good alternatives to live
foods are not yet available.

Harvesting has not been a problem since lobsters are
usually reared in tanks. Processing methods are efficient
and need little modification.

Potential for Culture of Other Crustacea

Several kinds of crabs, pandalid shrimp, Australian
crayfish, and other crustaceans have been suggested as
suitable candidates for aquaculture. Spiny lobsters have an
even more complex and prolonged larval life than northern
lobsters, and the research on raising them is still in early
laboratory stages. Commercial systems will probably not be
developed for many years. The culture of crabs has, thus
far, consisted primarily of raising larvae in laboratories.
Crabs grow slowly and require expensive animal foods.

Although some of these species may be good candidates,
so little information is available concerning their
biological requirements that their aquaculture potential
cannot be evaluated at the present time.

Marine Finfish

The availability of many desirable species of marine food fish will probably continue to decline as long as wild stocks remain the principal source of supply. The need for marine fish aquaculture may, therefore, increase. Unfortunately, many complex biological and technological problems require solutions before aquaculture can provide reliable alternative sources of marine food fish.

In Southern Asia, finfish, principally milkfish and mullet, account for more than half the aquaculture production. Several kinds of flatfish have been successfully cultured experimentally in Great Britain and France but the establishment of commercial industries has been delayed by rising costs. Japanese fish farmers have established profitable culture systems for a number of marine fish species, including sea bream and yellowtail. Commercial attempts to raise pompano in the United States showed considerable promise in the 1970s. In Florida, wild-caught juveniles were raised in ponds and tanks. Subsequently, the state prohibited capture operations, and workers perfected seed production by hormone-induced ovulation. Larval rearing problems coupled with disease and market uncertainties eventually forced abandonment of commercial efforts. Although the technology is available to raise mullet, the market for this species is currently too weak to justify its commercial culture in the United States. A tropical species, the rabbitfish (_Siganus_), shows some promise in Palau and Guam.

The many species of marine fish that can potentially be cultured will require production systems that are appropriate to their individual needs, but some generalizations can be made.

The technology exists for seed production of marine finfish either with the use of hormones or through temperature and light manipulation. Nevertheless, problems in rearing production quantities of finfish having minute eggs persist.

During larval stages, marine fish are typically fed plankton. Food requirements of larvae are fairly well understood, but technology for commercial production of larvae is generally undeveloped.

Although problems associated with growing juvenile marine fishes in captivity are often less difficult to solve than problems associated with larvae, information is needed pertaining to nutrition, disease control, and environmental requirements of marine fish. Considerable research and development work is required before commercial-scale

aquaculture can become a reality with the great majority of
candidate species.

There is only modest experience in this country in
rearing marine finfish. These species include red drum,
spotted sea trout, black drum, dolphin, mangrove snapper,
and yellowtail snapper. Technological problems could
probably be overcome if a strong market could be identified.

POTENTIAL PRODUCTION SYSTEMS

To date, most U.S. aquaculture has been devoted to luxury
food species that are, for the most part, high on the food
chain, fed prepared diets high in animal protein, and grown
in intensive production systems. Success with this approach
has, at best, been marginal, primarily for economic reasons.
Advancement in such fields as nutrition, genetics, and
disease control can undoubtedly improve the economic
picture.

The rapidly changing situation in the United States with
respect to cost, availability, and acceptability of aquatic
foods appears to justify the parallel development of new
initiatives in the field of aquaculture, involving concepts,
approaches, and species not yet utilized on a commercial
basis in this country. These new programs should include:
(1) cultivation of aquatic plants for food, feed, chemical
products, waste treatment, and biomass for conversion to
energy; (2) closed-cycle aquaculture; (3) polyculture; (4)
ocean ranching; and (5) alternatives to current feeding
practices. Such programs in aquaculture are in many cases
interrelated and even interdependent, in some cases
requiring simultaneous or at least intimately sequential
development. None is new or untried on a worldwide basis,
but neither has any of them been attempted at the high level
of intensity, technology, and sophistication that appear to
be the essential ingredients of economically viable
aquaculture systems in the United States. Research is
therefore needed, not only in the basic scientific
disciplines involved in the cultivation of the organisms,
but also in the engineering design and evaluation of high-
intensity culture systems that are both technically reliable
and economically viable. Demonstration facilities will be
required.

Aquatic Plants

Substantial quantities of seaweed are consumed in Asian
countries. In fact, seaweed production is the largest
component of the aquaculture industry in Japan. No such
seaweed food industry exists in the United States, and

unless food habits change or an export market is developed,
one is not likely to be created.

Seaweeds possess substances that can be extracted for
use as gelling agents. In the United States, although small
quantitites are harvested for this purpose, kelp is the only
seaweed that is currently being cultured. To increase
supplies of extractives and to produce methane for fuel,
systems are being designed for large-scale transplanting and
culture of kelp and other seaweeds. If demand for these
products continues, non-food marine plant farming may become
a significant industry.

Many of the problems that have plagued the development
of aquaculture in the United States are not applicable to
aquatic plants. In fact, some of those problems might be
alleviated through wise use of aquatic plants.

Aquatic plants include both algae and the higher,
flowering forms of vegetation. Among the former are the
unicellular forms (phytoplankton), filamentous species, and
the larger, macroscopic algae or seaweeds that exist almost
exclusively in saltwater. The higher forms of plants are
more common in freshwater where there are over 150 species,
but also occur in brackish and saline waters. All of these
forms of aquatic plants have existing or potential uses. In
some cases, the uses are specific to the species; in other
cases, the uses are common to all plants, although some
plants may be used more efficiently than others depending on
their rates of photosynthesis, growth, and nutrient
assimilation.

All autotrophic (photosynthetic) plants remove dissolved
nutrients, phosphorus, iron, and other minerals from the
water. These plants could be used to carry out a form of
advanced treatment on various kinds of wastewater. Such
treatment, if done in a highly concentrated plant culture
and under controlled conditions, may prevent eutrophication
of the aquatic environment.

In the future, aquatic plants may be cultivated as an
energy source. These plants, through digestion and
fermentation, could produce methane or other fuels or fuel-
sparing products. A new technology might be developed for
the construction and operation of aquatic energy farms.

Closed Cycle Aquaculture

People have practiced various forms of aquaculture for
thousands of years. While efforts have been directed toward
manipulation of natural systems, experience in agriculture

has shown that certain crops are best produced in artificial
or otherwise highly controlled environments.

Aquaculture food factories offer a means for controlling
production to a degree that would be impossible in nature:
(1) output can be adjusted to demand; (2) yield, size,
shape, texture, and contamination by pathogens may be
controlled; and (3) inorganic chemicals can be converted
into food more efficiently when solar energy is used under
controlled conditions. These factors suggest that
aquaculture production from closed systems may become a
profitable industry in the near future.

Closed systems have now been developed and operated at
the research level for salmonids and other finfish such as
yellow perch, penaeid and freshwater shrimp, and oysters.

Oysters were selected as suitable subjects for
demonstration for a number of reasons: (1) the products
have a high unit market value; (2) the decline in the
natural resource does not appear temporary; (3) the research
and development for aquaculture of oysters over the past few
years has had positive results; and (4) oysters are
efficient converters of algae into meat. The system for
oysters that has been operated at the research level
includes production of unicellular algal food. These algae
efficiently convert solar energy into food for oysters.
Partial recycling of water is efficient because the wastes
of oysters are nutrients for algae.

The oyster culture plant can be described as a partially
recycled, environmentally controlled algae/oyster husbandry
system, which is supplied with inorganic nutrients and
driven by solar energy to produce food for human
consumption. It is now at a pre-production prototype stage.
Cooperative inputs from industry and academia are needed to
bring it to commercial fruition.

Although aquaculture food factories offer considerable
potential, numerous problems must be solved before these
systems will be widely used commercially. The systems are
energy intensive, requiring considerable energy for
circulation and aeration of the water and for handling of
the animals. The expense of closed systems requires that
systems operate under maximum capacity to be profitable.
Therefore, maintenance of the life-support systems is
essential. If a critical component of the system fails and
is not repaired, the whole crop may be lost within as little
time as 30 minutes. Because such repairs may not be
possible, back-up systems must be available. These
additional systems add considerably to production costs.

Theoretically, closed-cycle aquaculture systems may be located without regard to the external environment. Currently, systems require a supply of make-up water amounting to as much as 10 percent of the total culture volume per day, and geographic location is therefore still important. Disease and pollution, if accidently introduced to animals raised in closed systems, may mean destruction of the entire crop. With adequate precaution, contaminants such as chemicals and pathogens can be avoided or treated. Prevention of toxic effects from metabolites and decomposition products of the cultured animals themselves and of their food require sophisticated and reliable treatment processes as an integral part of the system. Constant monitoring and control are essential if closed systems are to be free from environmental hazards.

Polyculture

Chinese aquaculturists have long recognized that yields can be greatly increased by stocking ponds with several species that occupy different ecological niches. The classic example is Chinese carps, of which only the grass carp is fed agricultural wastes or kitchen refuse, and four or more other species subsist on the undigested fecal matter and the planktonic and benthic communities of organisms supported by the residues of the herbivore.

Culturists in other countries have applied the techniques of polyculture to other combinations of species including: oysters and lobsters, oysters and rabbitfish, grass carp and freshwater shrimp, catfish and tilapia. Research on polyculture has just begun in the United States. The yields and efficiencies of various combinations of organisms need to be investigated, including combination of food organisms and target species that are grown together in aquaculture systems.

The intensive cultivation of finfish and invertebrates high on the food chain has, in some cases, failed in the United States for economic reasons. Because food usually represents about half the operating costs of such cultivation, reduction or elimination of those costs could place aquaculture in a new financial perspective. Unfortunately, this is not a practical solution for the culture of most of the luxury species that are presently in greatest demand in the United States. However, as supplies decline and costs escalate for the more familiar species, demands for aquatic food organisms are changing rapidly. The time is, therefore, ripe for research and development in the culture of aquatic organisms lower on the food chain, including both plants and the animals that eat them, which

can be grown in multi-purpose, semi-natural aquaculture
systems.

Fish that have shown great promise in polyculture
experimentation in the southern United States have been the
tilapia, silver carp, and grass carp. These fish, feeding
in niches heretofore unused in catfish ponds, increased
total yields 79 percent, to 4,300 kg/ha, without extra feed
or fertilizer. Multiple cropping of fish, such as raising
catfish in the summer and trout in the winter, should also
be investigated.

Ocean Ranching

In the practice of ocean ranching, juvenile anadromous fish
(in particular, Pacific salmon) are released to the sea
where, feeding on natural foods, they develop to adulthood
and then return to their home stream to spawn. In
principle, ocean ranching includes the planting of any
juvenile organisms in the environment from which they obtain
their food. Conventional methods of the culture of oysters
and other molluscs are examples of ocean ranching. The
aquaculturist strives to retain ownership or control of at
least a portion of the returning or remaining population.
The practice has proved successful both in the United States
and elsewhere. However, ocean ranching is handicapped by
lack of knowledge concerning the optimum time and the
optimum size at which fish should be released at different
geographic locations.

Between the extremes of organisms that return to the
culturist by natural or manipulated behavioral responses and
those that are immobile and remain where they are released,
there is another large, undeveloped method of ocean ranching
that merits investigation and evaluation. The animals are
not confined within and do not return to a specific harvest
site, but the hope is that they will remain reasonably near
their release point and enhance natural populations so that
conventional fishing may produce greater yields. The
Japanese have been leaders in this area, releasing to their
coastal waters a variety of juvenile marine organisms
including penaeid shrimp, abalone, scallops, and certain
finfish.

This form of ocean ranching is similar to the old and
long-abandoned practice of releasing hatchery-reared fish
that was a major undertaking of the (old) Bureau of
Commercial Fisheries. Modern efforts differ in that
organisms now stocked are hatchery-reared juveniles already
developed beyond the stages at which they suffer their
greatest mortality in the wild. This approach is appealing
because it takes advantage of the relatively low costs of

rearing juveniles in hatcheries, but avoids the high costs of rearing them to market size.

Success of the Japanese experiment is yet to be shown. The social, legal, and environmental constraints on such forms of ocean ranching are considerable. Aquaculturists, to obtain benefits from ocean ranching, must be able to retain control of their crop. New experimental programs involving all aspects of modern ocean ranching would appear to be warranted at this time.

Alternatives to Current Feeding Practices

Finfish culture in the United States has concentrated on species relatively high on the food chain that require costly feeds containing a significant portion of animal protein. In contrast, in China and Southeast Asia, herbivorous and omnivorous species are cultivated that derive much, if not all, of their diet from living or dead plant material or other forms of organic detritus. The more common, non-intensive pond culture practices normally used to grow these fish depend upon natural foods, assisted at most by fertilization to enhance plant production. However, these same fish may be grown intensively by means of heavy feeding with a wide variety of agricultural wastes and residues, and most species in question will readily accept pelletized artificial feeds.

A United States market for herbivorous fish is almost non-existent at present, but eating habits and acceptability of new aquatic foods are changing as conventional species become increasingly scarce and prohibitively expensive. Efforts to increase knowledge of the nutritional requirements of fish low on the food chain and the technology of formulating low-cost artificial diets for them may deserve equal priority with efforts to develop suitable feeds for fish species currently produced commercially.

An alternative to the use of artificial, prepared feeds in intensive aquaculture practices and to the dependence upon natural food in extensive systems is the intermediate step of growing natural food organisms as a part of the aquaculture system, either separately or together with target species. This method is currently used in the larval rearing of certain species of finfish and invertebrates for which a suitable artificial feed is not yet available; it usually involves the cultivation of unicellular algae, rotifers, or brine shrimp. Even for that purpose, however, the technology for the mass cultivation of those and other planktonic organisms is not adequately developed, practices are empirical and unreliable, and the larval food supply is often a constraint to aquaculture.

Some target species for aquaculture depend upon small, planktonic organisms throughout their lives. Present culture techniques for oysters and other bivalves is actually more a form of conventional capture fishing than aquaculture; the molluscs depend entirely upon an unpredictable and unreliable supply of natural food organisms in their environment. Controlled aquaculture for these species must develop around a regulated source and supply of food, the most likely form at present being cultured unicellular algae. Much is known about the physiology, nutrition, and growth in culture of the planktonic algae as well as their suitability as food for bivalve molluscs. But little work has been done on mass culture of marine algae and its engineering and economic aspects.

Other existing or potential aquaculture species, such as the silver carp, also feed on unicellular algae. Some species, including particular salmonids, may be grown throughout their lives on small planktonic animals. "Browsers" subsist on filimentous algae, seaweeds, or the higher freshwater aquatic plants.

Aquaculture systems that recycle wastes have been tested, but engineering design and economic analysis are needed to put such systems into commercial application. Public health problems may be created by such systems if cultured organisms are not prevented from concentrating pathogens and toxic trace contaminants present in the wastewater. The possibility of such problems must be investigated, and solutions must be found if such problems do exist.

RECOMMENDATIONS

The aquaculture industry in the United States profitably produces a variety of species including: salmon, trout, catfish, crayfish, baitfish, oysters, and ornamentals. A large category of candidate species exists for which viable production systems have yet to be developed. Both groups are constrained by a number of problems relating to technology.

Constraints on production relate to such areas as genetics, reproduction, health maintenance, and water quality. But considering all of the species and all of the constraints, the primary production problems to be overcome relate to mortality of larvae and juveniles and to feed technology. In addition, production has been severely constrained because available technology has not been applied by user groups. We therefore recommend that primary consideration be given to:

- •   Improved technology for rearing cultured species during the early stages of their life cycles.

- •   Improved technology for formulating, processing, and delivering food and feed for cultured species.

- •   Construction of demonstration facilities.

CHAPTER II

SCIENCE AND TECHNOLOGY

INTRODUCTION

Agricultural technologies were first based on empirical
evidence derived from subsistence farming, rather than on
the rational and deliberate application of scientific
knowledge. As knowledge increased, the information was
organized to increase production efficiency and develop more
effective crop varieties and animal breeds. In recent
years, agribusinesses have emerged to manage complex systems
of agriculture with multidisciplinary components.

Systems development was largely accomplished at the
land-grant universities. Government funding, academic
excellence in sciences, and dissemination of knowledge
through extension services all contributed to the success of
modern agriculture. Contributions were also made by
government scientists and the industrial scientific
community.

Much of the scientific knowledge gathered for
agricultural purposes is relevant to aquaculture and
additional information has been generated by scientists
performing basic research in aquatic sciences. However,
aquaculture production differs from other animal production
systems in that many aquatic species we culture, or attempt
to culture, are carnivorous. Further, because the
activities of aquatic species and the factors that affect
aquatic species take place in a single medium (water), the
problems posed are different from those that confront
terrestrial animals that are affected by land and air, as
well as water. The current state of knowledge concerning
certain species of aquatic organisms is sufficient for
profitable production systems (see Chapter 1), but
additional information on nutrition, genetics, reproduction,
parasites, health maintenance, disease organisms, and seed
production would improve these systems. Such information
would also enhance the possibilities of developing
aquaculture for additional species.

A primary technical factor limiting aquaculture may be the inability of people to adequately control the complex ecosystems involved in aquaculture. Interactions of many complex variables such as water quality, soil characteristics, meteorology, and tides, as they influence growth and survival of the cultured species, create problems for which satisfactory solutions have not been developed. Management of ecosystems is made more complex by the multiple uses of environments, and the intrusion of pollutants. The problems of managing aquacultural ecosystems deserve high research priority.

If environmental limitations exist--if growing seasons are too short or water quality is poor--then further research in nutrition, genetics, and disease control may be inadequate to overcome problems arising from poor siting.

Although the primary constraints on current aquaculture systems are not scientific and technical, gains from scientific and technical research can contribute to improving the performance of the system and increasing the profitability of commercial ventures. If these benefits from research results are to be realized, increased communication must occur between scientists, engineers, ecologists, economists, and business managers.

NUTRITION

Fish are efficient converters of dietary calories into flesh (Reid 1975, Smith 1976). Unfortunately, the specific amount of nutrients required by fish and the amount of chemical nutrients available to fish from feedstuffs are only partially known. Practical diets, rather than being based on a least-cost nutrient balance formulation, are often formulated empirically. Specific nutrient requirements are known for only a few species reared under a few controlled environmental conditions; even these requirements do not take into account the effect of one nutrient on another during intensive aquaculture production. Economic success or failure of the aquaculture venture is directly dependent, however, upon the nutritional state of the stock being reared. Thus, effort needs to be directed toward obtaining more information on nutrient requirements and availability.

Nutritional Requirements

For any given species, major constraints exist in defining specific nutrient requirements as they relate to size, environmental temperature, and growing conditions. Much attention has focused on salmonids and specifically on the nutrient requirements of rainbow trout. These data have

been used, with marginal success, to formulate diets for
other species of fish reared under different environmental
and growth conditions. Experimental test diets to study
vitamin, amino acid, or fatty acid requirements have been
developed for rainbow trout, chinook, sockeye, and coho
salmon, and for channel catfish, common carp, red sea bream,
and the Japanese eel. These test diets have been used only
for very young fish raised under one or two specific
environments.

Clinical semi-microchemical techniques, used in other
fields of animal husbandry for deduction of nutritional
status, have not been developed for fish. Such techniques
would be invaluable in determining qualitative and
quantitative nutrient requirements for species that are
being considered for aquaculture.

The Commission on Animal Nutrition of the Union of
Nutritional Sciences (FAO 1976) has recognized that large-
scale development of aquaculture is constrained by
insufficient knowledge on quantitative nutrient requirements
of target production species. These constraints limit
orderly development of intensive aquaculture and affect the
success of extensive aquaculture ventures where
supplementary feeds or fertilizers must be used to ensure
good growth. As more information becomes available on the
specific quantitative nutrient requirements for each species
raised under a particular environment and culture technique,
economical rations can be formulated scientifically rather
than empirically. Such formulations will improve the
efficiency and effectiveness of aquaculture.

Larval Feeds

Practical methods must be developed to formulate an
effective food to rear larvae to a size that can consume
common, industrially processed fish and shellfish diets.
Most fish diets are manufactured by grinding, sieving,
pelleting, extruding, compressing, and crumbling the
ingredients and then drying or freezing the produced
mixture. Modern industrial equipment limits the fineness of
the feed produced; the diameter of small food particles
generally is not less than 100-200 microns (μ), far too
large for many larval forms to ingest and use.

Several other practical problems must be solved before a
reliable larval food can be produced. A larval feed must
be: (1) water insoluble; (2) approximately 1.0 in density;
(3) palatable; (4) digestible; (5) satisfactorily balanced
to meet the nutrient requirements of the species; and (6) no
greater than 50 μ in particle size. A new technology must
therefore be developed, integrating the techniques of feed

science engineering, nutrition, and commodity processing.
Production of larval feed requires a fundamentally different
approach to ensure inter-nutrient component binding in order
to produce a stable compound that is nutritionally adequate
and that is capable of being micropulverized and separated
into stable larval food.

Fish Feedstuffs Nutrient Data Bank

Although the International Feedstuffs Nutrient Data Bank is
available for terrestrial animal husbandry, no comparable
data containing information on nutrients in fish feeds is
available for aquaculture species. Techniques have recently
been developed to measure the amount of energy available and
the amount used by fish in metabolic chambers (Smith 1971,
1976). The data produced by these techniques can be used to
predict, with a high degree of confidence, energy limited
growth in large-scale production systems. The techniques
have been used with rainbow trout, but have not been
extended to other species.

An international network of information has been created
that will be able to produce computer printout lists of
nutrient content. The system is ready to receive
information from cooperating agencies. Data on the
digestibility, availability, and use of nutrients will allow
least-cost computer formulation for each particular
aquaculture application. The primary constraint for current
use of this system is the lack of data on fish feedstuffs in
the computer storage system. As input increases, the
reliability, effectiveness, and efficiency of the data bank
will also increase.

GENETICS AND BREEDING

The application of genetic breeding followed by good
management and adequate disease control has greatly inceased
the efficiency of livestock production. The application of
genetic principles to breeding stock selection in commercial
or public aquaculture lags far behind in all technologically
advanced countries. For example, although Japan is heavily
dependent on fish (a significant portion of which is
cultured) and has a strong National Institute of Genetics
and excellently trained geneticists, only recently has Japan
begun to employ genetics in aquaculture.

Genetics

The basic genetics of fish in general is poorly understood.
In 1962, the American Fisheries Society recognized this fact

and urged the inauguration and support of research programs
on fish genetics by academic institutions and government
agencies.

United States programs of aquaculture genetics and
selective breeding typically lack the magnitude and
continuity necessary for studying organisms with relatively
long life cycles whose generation-to-generation maintenance
poses special problems. Genetic improvement of stock often
requires a considerable investment of time and money before
improved strains are available.

In view of the important role genetics has played in
improving crops and livestock, little doubt exists that this
discipline can be of value to aquaculture. Although the
state of aquaculture genetics is still quite primitive, some
work has been done. Genetics is intimately interwoven with
disease resistance, pathology, and nutritional requirements.
As a consequence, agricultural successes dependent on
genetics have been so related to simultaneous advances in
nutrition, pathology, and fertilizers that the specific
contributions of genetics are not easily separated from
those of other disciplines.

The failure of aquaculture genetics to contribute more
to commercial productivity is due to the relatively
undeveloped state of technology within the aquaculture
industry itself.

Genetics and breeding have costs that go beyond the
actual costs of conducting genetic research or breeding
programs. As aquaculture organisms lose those gene
combinations that adapt them to life in the wild and as they
are pushed by artificial selection and other genetic methods
to artificial peaks of productivity, their demands for food
and environmental control become greater.

Breeding

Approaches used in breeding programs intended for intensive
aquaculture, where the organism lives in a highly controlled
environment throughout all or part of its life cycle, should
differ from the approaches used for extensive aquaculture,
where the organism may be returned to nature for part of its
life cycle. For extensive aquaculture, emphasis should be
placed on the conservation of natural genetic variability;
the role of superior hybrids will probably be more prominent
than in intensive selection. The high productivity demanded
by intensive culture will, on the other hand, necessitate
selection that narrows genetic variability.

Genetic breeding in extensive aquaculture has caused legitimate concern among fishery management specialists who fear that plantings of hatchery fish whose genetic makeup has been altered may affect the native stock. Native salmonids are of particular concern. Although native fish may be more adapted to a particular area than hatchery fish, the native stock may be potentially endangered by hatchery plantings through hybridization of native and hatchery fish resulting in disruption of co-adapted gene pools. They may also be endangered through competition for spawning and rearing grounds and by such factors as earlier hatching of the progeny of hatchery fish.

Because many aquaculturists must depend on wild animals for breeding stock, they are reluctant to permit introductions of related species. Without such introductions, productivity is not apt to be increased to the levels necessary for some intensive aquaculture ventures to become commercially successful. In agriculture, for example, few of our major species are native to the United States. The systematic exploration and testing of wild genetic resources is one of the first steps in establishing any intensive breeding or stocking program.

Exploitation and domestication of wild stock and environmental degradation will result in irrevocable loss of genetic diversity. Selection response depends on genetic heterogeneity. If aquaculture is to be stimulated, some coherent policy on stock conservation (wild and hatchery) must be established to maintain genetic diversity for present and future needs of aquaculture.

Domestication and intensive aquaculture, particularly monoculture, will intensify disease problems. Breeding for disease resistance, therefore, is an area of central importance.

The production of improved stocks for aquaculture, almost irrespective of species, should proceed as follows: (1) elucidation of the genetic structure of the wild populations; (2) domestication of wild stocks to create genetic types more generally amenable to aquaculture; and (3) improvement of strains of domesticated stocks involving selection, hybridization, and, perhaps, occasional deliberate inbreeding.

Attractive as probable gains of selection programs may be, breeding improvement of aquaculture organisms can be successful only if the aims of the selective breeding programs are clearly defined. New breeds alone are not likely to establish a business and create a market.

Hybridization can be used to combine desirable traits of
two different types, to create an entirely new type able to
live under circumstances neither parent could survive, or to
develop potential hybrid vigor. Some fish hybrids are
sterile and may have use in special production systems.

Research should also be conducted on induced
parthenogenesis and self-crossing of hermaphrodites to
rapidly develop a group of highly inbred animals for
experimental breeding.

Because aquatic species spawn large numbers of gametes
that are externally fertilized, cryopreservation of the
eggs, sperm, and perhaps even the already fertilized egg may
be possible. Although considerable research would have to
be performed to make this possibility a reality, the benefit
to the aquaculture industry could be large.

Cryopreservation would permit year-round availability of
seed stock and would make possible controlled and repeated
use of superior-performing individuals or strains.
Introduction of new stock via frozen gametes might minimize
the risk of introducing pathogens and would, to some degree,
facilitate conservation of various wild and domestic gene
pools.

Breeding programs require a span of years for continuous
generation-to-generation breeding. Such a program would
require a network of hatcheries in areas suitable for
culturing the important commercial species. Furthermore,
adequate provision would have to be made for field testing
of stocks. (See Appendix B for a discussion of the genetics
for important aquaculture species.)

SEED PRODUCTION AND AVAILABILITY

A major deterrent to successful culture of many species in
the United States is the unavailability of seed. However,
techniques for seed production, at least in finfish and
certain bivalves and crustaceans, are reasonably well known.
Because rainbow trout have been cultured in the United
States for nearly a century, the seed is available and in
demand. Commerce in eggs and fingerlings of rainbow trout
and other salmonids is well established.

The situation is similar for channel catfish, although
this industry is much younger than the trout industry. Both
trout and catfish produce large eggs that hatch into
physiologically advanced larvae. Advanced larvae can eat
and digest a variety of artificially prepared foods. Trout
and catfish also naturally reproduce in artificial hatchery
situations.

Other finfish species also reproduce naturally in
captivity, but produce small eggs and larvae. Such larvae
can only be reared in production quantities with natural
foods. Centrarchids and cichlids are examples of such
species. More often than not, a sojourn in earthen ponds is
a vital part of the culture of these fishes. A small
commerce has developed for such species, but seed is
expensive and only marginally available.

Still other finfish species must be encouraged to spawn
in captivity through use of special techniques, such as
hormone injection or temperature and light manipulation.
Such species usually produce very small eggs and larvae and
must be fed natural foods for several weeks after hatching
(most of the commercially interesting crustaceans possess
characteristics similar to this category of finfish). Seed
of such crustaceans is essentially not available to
commercial aquaculturists. Finfish in the last category can
be induced to ovulate in captivity by manipulating
temperature and light regimes in order to trigger the
release of endogenous gonadotrophic hormones. The process
is time-consuming and requires much equipment, human labor,
and skill. Broodstock are taken from the wild and
acclimated to laboratory conditions over a period of weeks
or months, where both the daylight and the temperature are
gradually adjusted to simulate natural spawning conditions.
Spawning usually occurs several weeks later.

Wild brood fish can also be spawned with injections of
gonadotrophic hormones such as chorionic gonadotrophin or
dried carp pituitaries. In order to produce a large number
of viable eggs, the fish must be physiologically within a
week or two of spawning. The fish are held in captivity for
several days until ovulation occurs. The sex products are
stripped and fertilized and incubation of eggs is achieved
using standard equipment and techniques.

The location and capture of eligible broodstock is
frequently difficult and expensive. Commercial fish
farmers, in many cases, are legally proscribed from
capturing eligible broodstock and usually have little
expertise in the use of hormones, temperature and light
manipulation, and artificial spawning and hatching. Fish
farmers have little incentive to gain these skills prior to
the determination that the candidate species can be cultured
and sold at a profit. Thus, the techniques for seed
production of many freshwater and marine species are known,
but the supply of seed and broodstock, except for salmonids
and catfish, is essentially unavailable to commercial
aquaculturists. The federal government should encourage
farming of new species by providing seed until the
feasibility and production can be demonstrated, and the

industry has the incentive to provide the service for
itself.

DISEASE

Losses due to disease, whether by slow continuous attrition
or by sudden catastrophic epizootics, are familiar problems
for aquaculture. Technology to reduce the impact of disease
is developing, but newer and often more difficult problems
arise when current problems are solved. Nevertheless, a
substantial body of practical health measures for
aquaculture has been developed. Many of today's
aquacultural health problems derive from ignorance that
results in violations of basic health principles.

Infectious diseases involve not only the pathogen and
its host, but also the interplay of the nutrition,
physiology, and genetic constitution of the host with a wide
array of environmental factors. Maintenance of aquatic
animal health means effective and efficient health
management of large populations. Measures that avoid or
minimize disease offer increased margins of profit and at
times can mean the difference between success and failure.

The study of disease in freshwater aquaculture has been
concerned primarily with salmonids and catfish. Attention
is currently focused on virus diseases of freshwater fish,
whereas a few decades ago, the bacterial diseases dominated
the attention of researchers.

Because intensive marine and estuarine culture has a
shorter history than freshwater culture, the state of
knowledge about marine diseases is correspondingly less
advanced. Major disease-caused mortalities have occurred in
cultured oyster populations both in the United States and
elsewhere (Sindermann 1976), and disease has contributed as
a deterrent to successful culture of salmon and shrimp in
salt water (Fryer et al. 1976, Johnson 1975). Knowledge
about marine diseases has advanced within the past decade,
especially in the critical area of larval diseases
(Sindermann 1974).

Health Maintenance

The technology of disease control has necessarily paralleled
the development of fish culture in freshwater. Beginning
with the simpler chemical measures (formalin, copper,
sulfate, malachite green), treatments have progressed
through the sulfa drugs to antibiotics and the nitrofurans.
Control of disease should concentrate on prophylaxis rather
than treatment--and prophylaxis must emphasize adequate

47

water quality and nutrition, as well as reduction of
stresses on cultured populations. Chemical prophylaxis (and
probably chemotherapy) should be considered as "last resort"
methods in disease control (Herman 1970).

A substantial body of practical health measures has been
developed for freshwater aquaculture. Although our
knowledge of prophylaxis and treatment of diseases in marine
aquaculture has increased, it is far from complete. Some
remedies developed for control of freshwater fish diseases
can be adapted for use with marine fish, but the field of
marine invertebrate disease control requires extensive
experimentation. Health maintenance depends on:

A.   <u>Preventive</u> measures, including:

   (1)   maintaining water quality;

   (2)   reducing environmental stress (low O_2,
temperature extremes, build-up of waste products);

   (3)   providing adequate nutrition;

   (4)   developing specific pathogen free and disease-
resistant stocks;

   (5)   developing vaccine for immunological
protection;

   (6)   manipulating environmental conditions (for
example, growing oysters in water which has salinity
levels that are too low for survival of pathogens);

   (7)   enforcing regulations that prevent national or
international transfer of pathogens from one host
population to another; and

   (8)   providing chemical prophylaxis.

B.   Correct <u>diagnosis</u> (including understanding the life
cycle and ecology of the pathogen) is a critical step in any
control program.

C.   <u>Treatment</u>, usually consisting of chemotherapy,
possibly combined with some preventive measures.

Chemicals added to culture systems for disease control
may serve two important functions: (1) they may reduce
pathogens; and (2) they may reduce or control populations of
heterotrophic microorganisms. The principal problems
encountered with use of chemicals or antibiotics include:

A.   Possible negative effects on biological filters in
controlled recirculated systems--particularly on nitrifying
bacteria;

B.   Possible negative effects on algal food, or on
algae present in fish larval rearing tanks; and

C.   Possible undesirable or harmful residues in
cultured animals.

A wide range of chemicals has been used to control fish
diseases. Sulfas and antibiotics have both had continued
use. At present, interest has focused on the nitrofurans--a
class of chemotherapeutics first used by the Japanese
against bacterial fish diseases. The development of drug
resistance has been and continues to be a significant
problem.

Chemical methods of disease control are severely
restricted when applied to animals being raised for food.
For example, only salt, acetic acid, and sulfamerazine are
currently approved by the Food and Drug Administration (FDA)
for use on all varieties of food fish. Oxytetracycline is
restricted to trout, salmon, and catfish. Applications of
such common and useful substances as formalin, copper
sulfate, acriflavin, and potassium permanganate are
currently illegal for treatment of species destined for
human consumption.

Furthermore, some chemicals may be carcinogenic or may
in other ways cause damage to humans who handle the
compounds. Some potentially harmful chemicals may actually
be used (even though they are not cleared) and such
chemicals may leave persistent residues in the harvested
product destined for human consumption. Other chemicals may
affect food-chain organisms in the natural environment;
their widespread use should be discouraged. Accelerated
clearance of chemicals, or publication of definitive data on
their beneficial or harmful effects, should have high
priority. The present status of registration of chemicals
for disease control in fish has been reviewed by Meyer et
al. (1976).

Effects of Transfers and Introductions

The introduction of aquatic species from other countries
poses specific threats of disease, particularly when the
introduction occurs in open-system culture. Introduced
species may be susceptible to enzootic pathogens, parasites,
commensals, and predators in new grow-out areas, and the
native species may be susceptible to pathogens, parasites,
commensals, and introduced predators.

Because transfers and introductions are likely to
continue, a plan of action should be developed to limit
risks of disease. Before introduction is permitted, a
detailed study should be conducted in the native habitat of
the organisms, concentrating on possible microbial diseases.
Quick parasite surveys should not be considered adequate.

The species to be introduced should be examined in
closed or controlled systems for an extended period--up to a
year--to see if disease problems emerge. Larvae should be
removed from any contact with brood stock; only offspring
should be permitted to be placed in open waters. Brood
stocks should be developed from these offspring, and
original stocks should be destroyed (disease can become a
major problem in maintenance of brood stock--whether
introduced or not--and transmission of disease to eggs or
larvae should be prevented).

Effective quarantine needs greater emphasis. Initial
introduction of small numbers of offspring of foreign
species should not be done in open waters without thorough
prior study and North America quarantine. Adequate
quarantine could have prevented problems such as the
introductions of the molluscan oyster drill and the cyprinid
fish tapeworm recently brought into North America with the
Asiatic grass carp.

Quarantine measures should be integrated with
inspections and certifications of viable entities from
foreign sources. The Lacey Act (50 CFR 16 [18 USC 42]) was
enacted to protect natural U.S. stocks from introductions of
diseases and to prohibit introduction of particular species
whose feeding habits represent a threat to existing species.
Had this legislation been enacted sooner, the introduction
of the European myxosporidan parasite of salmonids might
have been prevented. The Lacey Act requires that imported
salmonids be certified to be free of European Viral
Hemorrhagic Septicemia--a devastating disease of young and
adult trout. Thus far, North America is free of the causal
virus, in large measure because the United States and Canada
both require certification.

ENVIRONMENTAL POLLUTION

The edges of the sea cannot be used simultaneously as
cesspools and as producers of food. When particular
combinations of toxic chemicals, low oxygen levels, and high
temperatures occur, oysters, clams and certain finfish
disappear from areas that formerly had abundant fish. Such
was the case in Raritan Bay in New Jersey and Escambia Bay
in Florida.

Effects of contaminants in less polluted bays,
estuaries, and coastal waters are less obvious. Marginal
environmental conditions caused by toxic chemicals, low
oxygen, and diminished populations of certain forage
organisms may produce stresses on fish and shellfish. These
stresses result in slow growth, greater susceptibility to
infectious diseases, reduced survival of young, and a
gradual diminishing of species important to recreational or
commercial fishermen.

Research to date has not demonstrated that pollutants in
estuaries and coastal waters are entirely incompatible with
open-system aquaculture, but it has indicated the possible
severity of the problem (Sindermann 1970, 1972). The major
contaminant-related problems in open-system aquaculture
appear to be those produced by industrial sources or
domestic wastes.

Public Health Problems

Ocean pollution that affects aquaculture can affect public
health by microbial or chemical contamination of food.
Except for a few well-publicized chemical problems, such as
Minamata Disease in Japan and high mercury levels in
swordfish, available information primarily deals with
microbial contaminants.

Viral and Bacterial Contaminants

Microbial problems related to human public health can be a
major deterrent to open-system aquaculture. Virus diseases,
such as hepatitis, have been shown to be transmitted by
ingestion of raw shellfish from polluted waters (Mason and
McLean 1962). Experiments indicate that the survival time
of viruses in salt water varies but can be surprisingly long
(Metcalf and Stiles 1966). Limited information is available
about the viability of viruses in sludges, bottom sediments,
and marine organisms.

Thus, serious problems confront aquaculture operations
that are located in areas where even limited domestic
pollution exists if such operations produce food that is
consumed by humans in a raw or partially processed state.
Most areas currently used for aquaculture suffer from such
pollution. Viral contamination will remain a problem
wherever treated waste products are used for enrichment of
growing areas, unless inexpensive techniques are developed
that can absolutely purify the product or destroy the virus.
The danger to public health in the United States from fecal
contamination of growing areas in other countries has
increased with our increased importation of seafood. To

minimize this danger, FDA requires data, such as coliform
counts, in areas from which shellfish are exported. FDA
also sends research teams to evaluate risks of
contamination.

Although viruses constitute the most vexing public
health problem in open system aquaculture, pathogenic
enteric bacteria also pose a threat. At present, attention
is being paid to the role of a marine vibrio (<u>Vibrio
parahemolyticus</u>) in outbreaks of "food poisoning" in the
Orient and more recently in the United States. Although
this vibrio is a normal constituent of the inshore flora,
its abundance may sometimes be increased by the organic
enrichment of coastal and estuarine areas. Other pollution-
associated bacteria, such as <u>Clostridium</u>, <u>Salmonella</u>, and
<u>Shigella</u>, should not be ignored; even a single outbreak of
disease related to any marine species can have drastic
impact on markets for all marine products.

Chemical Contaminants

Chemicals can accumulate in seafood at levels that are toxic
to humans. Because of the horrors of Minamata Disease,
caused by mercury contamination of fish and cultivated
shellfish in a bay in Japan, mercury has received more
attention than other similar chemicals. While increased
surveillance of heavy metals in seafood has reduced
likelihood of another Minamata incident, risk of chemical
contamination exists wherever food is grown near industrial
operations. Furthermore, contamination is possible because
surveillance methods cannot easily be extended to include
every locality, and the toxic levels of many contaminants
are not yet known.

Certain processing methods, particularly the production
of fish protein concentrate, result in relatively higher
concentrations of mercury in the product than in the raw
fish. These concentrations may exceed allowable limits for
human consumption (Landis 1972). Removal of mercury (and
presumably of other contaminants) is technically feasible
(Regier 1972) although the economics of removal may preclude
its application.

The possible toxic effects of pesticides in foods have
received much attention. In 1965, a monitoring program was
instituted in the United States to scrutinize pesticide
levels in estuarine fish and shellfish (Butler 1969a). Thus
far, pesticide residues in all samples from the 15 states
involved in the program have been below levels considered
hazardous to humans. Although indiscriminate use of
pesticides is coming under some measure of control in the
United States, their continued use, persistence in the

environment, and accumulation through successive stages of
the food chain suggest that pesticides pose a potential
threat to nearshore ocean areas devoted to aquaculture.

Polychlorinated biphenyls (PCBs) and a host of other
organic trace contaminants in industrial wastes are now
widespread in the aquatic environment. Many of these
compounds are similar to pesticides in that they are long-
lived, accumulate in successive stages of the food chain and
are toxic to cultivated organisms or humans.

Because some petroleum derivatives can be carcinogenic,
opposition has developed to undersea oil drilling and
offshore oil terminals (see, for example, Grossman [1972]).
Those who are opposed note that oil is persistent for long
periods in the marine environment, and that oil hydrocarbons
are taken up by organisms in the food chain and transferred
to humans when they consume seafood.

Natural biotoxins also pose public health problems, in
particular, paralytic shellfish poisoning. Some molluscan
species from certain coastal areas have had toxic effects on
humans because the animals had ingested and accumulated
toxic microorganisms. Some species of fish from particular
locations in tropical and subtropical waters have long been
known to be toxic, but the etiology of their contamination
remains uncertain. Recently, changes in the distribution of
the the dinoflagellate _Gonyaulax_ _tamarensis,_ which causes
paralytic shellfish poisoning, have occurred. These changes
in distribution may be related to changes in nutrient levels
caused by anthropogenic pollution of the environment. In
the summer of 1972, for example, on the New England coast,
blooms of _Gonyaulax_ were noted further south than in prior
years and caused toxicity in clams and other bivalves.
Outbreaks in Europe appear to be related to very high
nutrient levels in coastal and harbor areas (Korringa 1968).

For ciguatera, poisoning caused by eating species
containing ciguatoxins, Halstead (1971) stated "the field
evidence...strongly suggests that under certain
environmental conditions in tropical insular areas,
pollutants may provide the necessary chemical constituents
to trigger 'naturally-occurring' biotoxicity cycles such as
ciguatera...." Fish poisoning is an important economic and
public health problem in the Caribbean and in tropical
Pacific islands. Possible pollution-mediated biotoxins
should be an important consideration in planning aquaculture
operations, particularly (but not exclusively) in tropical
waters.

Reproduction, Survival and Growth of Marine Animals

Anthropogenic pollution of a chemical, thermal, or
biological nature affects the environment of estuarine and
coastal species. Such pollution can place continued stress
on aquatic organisms and in some cases may lead to
destruction of certain species. Research on this topic is
important for the development of aquaculture.

Diseases of Fish and Shellfish Related to Pollutants

The rich organic soups created in inshore areas by domestic
sewage outfalls and sludge dumping can contribute to the
expansion and modification of the normal marine and
estuarine microbial flora. The organic content of water or
sediments is increased in most aquaculture areas and hence
provides opportunity for increases in populations of
facultative bacteria such as vibrios, pseudomonads, and
aeromonads. Concentrations of such bacteria may provide
sufficient infection pressure on fish or shellfish so that
disease and mortality results. Evidence for this sequence
of events can be found in studies conducted in heavily
polluted waters (Mahoney 1970, American Littoral Society
1972) in which bacterial fin rot occurs in many species of
fish.

Some preliminary data suggest that some bacterial
pathogens of humans may be able to infect fish. Janssen and
Meyers (1968) demonstrated that fish taken from parts of
Chesapeake Bay near human population centers possessed
antibodies to a spectrum of human pathogens. Fish taken
from waters near sparsely populated areas did not contain
the antibodies (Janssen 1970). This topic needs to be
studied further. Although human pathogens may not be lethal
to fish, the presence of antibodies could be used as
indicators of pollution levels in aquaculture areas.

A particular microbial problem has emerged concerning
the use of heated effluents in aquaculture operations.
Fossil fuel and nuclear electric generating plants produce
significant amounts of heated water, which can be used for
year-round growth of aquaculture animals. Growth rates of
species such as American lobsters, oysters and catfish have
been found to be significantly greater in these elevated
water temperatures. However, some evidence suggests that
the positive effects of thermal additions may be offset by
increased mortality due to disease. Farley et al. (1972)
described a lethal virus disease of oysters in heated
discharge water in Maine. The disease, apparently low-level
enzootic in oysters growing at normal environmental
temperatures (12-18°C summer temperatures), has more serious
effects on oysters grown at elevated temperatures (28-30°C).

Studies at the Sandy Hook (New Jersey) Laboratory of the
Northeast Fisheries Center indicate a high occurrence of
lymphocystis disease in limited samples of striped bass
overwintering in the heated effluent of a Long Island
generating station. This virus disease is considered rare
in striped bass (Anonymous 1951, Krantz 1970), and its
frequent occurrence in a localized population may be related
to the abnormally high winter temperature of the water in
which the population exists. As with the oyster virus
disease, the high temperatures may promote survival or
transfer of the pathogen, or lower resistance of the host,
resulting in recognizable stages of infection. An
additional concern about fish overwintering in heated
effluents is that a source of infection will be provided to
incoming spring migrants. Bacterial fin rot of striped bass
overwintering in heated effluents has also been reported
recently by fishermen.

Effects of Chemical Pollutants

Chemical changes in the estuarine and nearshore waters are
of particular concern. Chemical pollutants are introduced
through pesticides in runoff from agricultural areas,
through heavy metals and PCBs from industrial effluents, and
through petroleum and its derivatives from polluted air as
well as from off-loading facilities and ship accidents.
Evidence is accumulating slowly about the acute effects of
high levels of such contaminants and about the chronic
effects of lower levels. Many parts of our oceans have
already been chemically modified by human activities.
Studies are necessary to discover the extent of change and
the tolerance of individual aquatic species to such changes.
The physiological and genetic reactions to environmental
degradation of the zygote, larvae and early juvenile stages
are of particular importance. Possible genetic effects on
subsequent generations should also be studied.

    The extreme sensitivity of molluscan larvae to
pollutants has been recognized and used in the development
of bioassay techniques, particularly on the west coast of
the United States (Woelke 1967, 1968). Additional evidence
of molluscan larval mortality after experimental exposure to
heavy metals has been obtained in Japan and Australia (Okubo
and Okubo 1972, Wisely and Blick 1967). Currently, effects
of heavy metals on embryos and larvae of oysters and clams
are being investigated at the Milford Laboratory of the
Northeast Fisheries Center. Their early findings indicate
that even low concentrations of certain metals may inhibit
development or cause death. Other related studies at
Milford concern possible effects of heavy metals on mitotic
rates, division irregularities, and chromosomal damage in
oyster embryos.

Effects of pesticides on the early life of aquaculture animals have long been a matter of concern. Davis (1961) and Davis and Hidu (1969) described severe retardation of embryonic and larval development in oysters and clams, as well as the lethal effects of 52 compounds (such as commercial pesticides and herbicides). It is likely that pesticides significantly increase mortalities of estuarine fish and shellfish during sensitive stages of their development. Specific levels of pesticides have been shown to slow the growth of adult oysters (Butler 1960), and residues found in oysters sampled from certain U.S. waters exceeded these levels. Furthermore, DDT-contaminated diets fed experimentally to shrimps, crabs, and fish (croaker and pinfish) caused significant and rapid mortalities (Butler 1969b). Effects of pesticides on reproduction of fish have been noted in other studies (Burdick et al 1964). PCBs are widely distributed in the marine environment, and recent studies (Risebrough 1969, Duke et al 1970) have demonstrated their toxicity to juvenile shrimp.

Petroleum products also pose chemical problems for aquaculture. Risk of massive contamination of growing areas by accidental spills will always be present, especially as ocean transport of petroleum increases. Long-term contamination of coastal waters by air fallout, pleasure boat exhausts, and sewage treatment plant discharges can affect growth and quality of marine animals. Studies at the University of Rhode Island (Anonymous 1972) indicate that hydrocarbon pollutants cause a "stress syndrome" in clams that include reduced growth rate, poor condition, shortened life span, and other physiological abnormalities. Sidhu et al. (1971) found that Australian mullet took up petroleum hydrocarbons from coastal waters with very low levels of pollution; fatty infiltration of livers and high lipid content of muscle characterized the affected fish. Fish eggs and larvae were found by Kuhnhold (1971) to be sensitive to various crude oils. Mortality of eggs was proportional to concentration, and larvae were more sensitive than embryos.

All these constraints suggest that if extensive coastal and brackish-water aquaculture is to be successful in the United States, the best estuarine and coastal grow-out areas should be protected from further encroachment by harmful chemical contaminants. The possiblity of aquaculture in interior saline lakes may also merit investigation. At the same time, research should be directed toward development of truly intensive closed-culture systems that recover waste products and recycle the water supply.

RECOMMENDATIONS

Knowledge in scientific disciplines has enabled successful
production systems to be developed for a few aquatic
species. Additional knowledge can improve the efficiency of
these systems and create a foundation for new ones.

   Such improvements will arise from basic research in
aquatic nutrient requirements, genetics, breeding,
reproduction, health maintenance, water quality, and
behavior. Specifically, we recommend:

- Research to determine the nutritional
requirements of culturable species throughout their
life cycles.

- Research to improve understanding of mass
culture in order to increase control over behavior
and reproductive processes so that the necessary
stocks of seed or juvenile organisms can be
developed.

- Research on the genetic characteristics of
culturable species in order to develop strains
possessing desired characteristics and to preserve
natural and domesticated brood stocks.

- Research to improve knowledge of disease
organisms and to develop appropriate disease
control techniques.

- Research to determine the effects of water
quality and the physical environment on cultured
species so that techniques may be developed to
maintain an optimal environment.

- Establishment of appropriate procedures for
registration of all exotic transfers for
aquaculture and for effective quarantine to prevent
introduction of disease.

CHAPTER III

ECONOMICS AND BUSINESS

INTRODUCTION

To be economically viable, an aquaculture venture must be
able to produce an aquatic biological product at a
competitive cost and be able to sell it at a reasonable
profit. Before such a state is reached, much time is
usually required for basic scientific research knowledge to
result in the development of improved technology and for
such technology to become commonly used within the industry.
In addition, new market avenues may have to be developed and
consumer acceptance may, in some cases, have to be
generated.

Economic issues are not easily separated from
scientific, technical, administrative and regulatory issues.
In this chapter, we address certain pertinent issues from an
economic point of view. Although the discussion of the
status of aquaculture may appear optimistic to some,
economic constraints definitely act as inhibitors to faster
growth. These constraints are categorized under the
headings of demand, production, marketing, and finance.
Throughout the discussion, we address issues pertaining to
property rights and public policy.

DEMAND

In economic terms, the demand for a product means the
relationships between quantities of that product that
consumers will purchase and such factors as the price of the
product, the level of consumer incomes, the prices of
competing (or complementary) products, and the size of the
consuming population. These relationships are determined by
consumer tastes and preferences.

Price of the Product

Analyses of the demand for most aquacultural products (and for most products) indicate an inverse relationship between demand and price; consumers purchase more of the product at lower prices than at higher prices. The extent to which prices would have to fall in order to increase consumption is important information for the aquacultural industry. If prices have to decline dramatically to increase consumption of a specific species (that is, if the demand is highly price-inelastic), aquacultural development of that species may be economically unviable.

Although growth of an industry generally implies expanded production, the profitablity of such expansion depends in part, on the resulting impact on price. For example, prices for finfish and shellfish that are considered "luxury" items are generally high and supplies are often short, but residual demand may also be limited: if increased production reduces total and net revenues, industry may attempt to limit the supply.

Demand for a product may be great enough to support a single aquaculture venture. However, if other producers enter the market, prices may fall and the industry may become unprofitable. The nature of demand coupled with production costs will determine the number of ventures that can survive in an aquaculture industry. In turn, the number of ventures may affect the degree of competition within the industry. Research to estimate the price elasticity of demand for various aquacultural species would be useful in predicting the viability of industries involved in production of those species.

Consumer Income Levels

Research on the relationship between demand for aquaculturally produced goods and consumer income levels has produced disparate results. Some empirical analyses indicate that within particular income ranges demand for aquacultural goods decreases as income levels increase. One explanation for this may be that consumers, as their incomes increase, substitute other foods (perhaps other seafoods) for the products formerly consumed, although the unit price of these substituted products may be higher. Estimates of demand for some species of canned salmon have been based on this assumption.

Although this relationship may not be characteristic of most seafood products currently produced in the United States, it is a potential constraint on the development of some kinds of aquaculture. Research in this area could be

59

useful in predicting the viability of industries based on
particular aquacultural products.

Prices of Other Foods

Demand for aquaculture products is also influenced by the
price of other foods. However, the products with which
aquacultural foods must compete have not been clearly
identified. Many products could be substitutes for oysters,
catfish, trout, salmon, or other fish foods, and these
products may substitute for each other. The actual
substitution made by a consumer may depend on a variety of
considerations including texture, versatility, and
nutritional content of the product.

The nature and degree of the substitution will also vary
according to the type of outlet at which the product is
sold: the product must compete with many types of food in a
supermarket, but only with other seafoods in a seafood
restaurant. Substitutions will also vary according to
whether the product is used as a final consumer good or as
an ingredient in the production of another good. Research
on the nature of these substitutional relationships will
help to determine whether aquacultural goods can compete
successfully with other products.

Number of Consumers

The size of the consuming population is an important factor
affecting the demand for aquacultural products. The
consuming population may increase naturally through
increases in the population, or through changes in the age,
race, and sex distribution of a given population. An
understanding of the role of these factors in consumer
demand for aquaculture products should assist in forecasting
the economic viability and explaining the growth of
aquacultural industries.

Producers may influence the size of the consuming
population through such actions as advertising and
exporting. As real incomes rise in other countries and as
territorial limits are expanded to further exclude foreign
fishing fleets, opportunities for sales of U.S. aquacultural
products in foreign markets may increase. Results from
research on substitutional relationships, price
elasticities, and other factors will improve estimates of
the potential size of foreign markets.

United States aquaculture industries must be sensitive
to the role of foreign aquacultural producers and must

realize that competition may take place in both domestic and
international markets.

PRODUCTION

The costs of aquaculture products depend not only on the
costs of the inputs to production but also on the
relationships between various input combinations and output;
in short, the production function for aquacultural species.
In this section we focus on the state-of-the-art with
respect to input-output relationships, organization forms of
aquacultural enterprises, and aspects of various markets for
inputs used in aquacultural production.

Specification of Production Relationships

In many instances, aquatic farmers have the knowledge to
produce a specified amount of food, but production is not
carried out in a technologically efficient manner. While it
is theoretically possible to decrease the use of one or more
inputs and still maintain production at its former level, or
to increase production without changing the input,
biological information on many aquatic species is inadequate
to develop a technologically efficient production
methodology. Until basic biological data are available for
projecting survival and growth rates of specific species,
production costs cannot be reliably estimated.

Although several combinations of inputs may produce the
same amount of food in a technologically efficient manner,
only one of these combinations will be economically
efficient, and that one will depend on the relative prices
of the inputs. Hence, different ways of producing the same
amounts of output must be studied so that economically
efficient methods of production can be established. Many
levels of output should be studied because the economically
efficient input ratio may be different depending on changes
in input prices and increases in production. Furthermore,
increasing production by a given amount may result in a
smaller increase in cost at a higher level of production
than at a lower level. Experiments in laboratories and
pilot plants may, therefore, not reveal the extent of costs
associated with larger scale production.

Economic Factors Affecting Hatchery Output

Costs of Feed

The largest cost item of production in many aquacultural
systems is feed. Feed costs in intensive fish production

61

amount to about 50 percent of total production costs. Small
changes in the price of feed and in the efficiency of food
conversion have large effects on the competitiveness of the
product. Animal protein and grains are major components of
most fish feeds and fluctuations in the commodity markets in
these items can cause serious dislocations. The higher cost
of fish meal after the recent Peruvian anchoveta failure is
an example. Use of animal and grain protein for aquaculture
may compete with agricultural uses and, in certain
instances, with human consumption.

An aquaculture feed must have good nutritional qualities
as well as low cost. Inadequate diets generally impair
growth, health, ability to survive, ability to convert
energy, appearance, and marketability, thereby reducing the
ability of aquacultural products to compete successfully
with other products. With regard to seed inputs, the high
capital costs of hatcheries and high productivities may
encourage development of a specialized sub-industry to
produce seed.

Losses from Disease and Predation

In some areas, hatchery-produced oysters have to be held for
a minimum of 29 months before they grow large enough for
market. During this period they obtain their food from
naturally produced plankton in open waters, but are still
subject to disease and predators. Any losses due to disease
and predators obviously affects the cost of the business.

Costs of Labor

Currently, most forms of aquaculture are labor intensive.
Oyster harvesting is still done largely by hand,
particularly in areas where state and local laws limit
harvesting from public beds to hand-operated tools. Within
the last two decades, however, substantial progress has been
made in reducing labor costs through mechanization. Due to
the relatively small size of the aquaculture industry,
companies have not been interested in designing and building
special aquacultural equipment. Progress has come through
adaptation of existing machines and a few new inventions,
for example, a mechanized boat that can harvest up to 2,000
bushels of oysters an hour with no more than four workers.

Energy has been receiving considerable attention
recently, but excessive concern with costs of energy may
result in false economies. The goal should be to minimize
total cost of production, not just energy cost. The optimal
input combination (i.e., the least cost) of certain methods
of aquaculture may involve considerable energy.

Optimum Production Methods

As animals increase in size, differences occur in feeding
rates, kinds of food, tolerance to disease, and other cost
factors. The most economical production technique may,
therefore, involve complicated patterns of stocking,
thinning, and moving individuals to different sized tanks
and ponds and the use of different feeds and feeding
patterns. The optimal pattern will also depend upon such
things as the nature of the processing and marketing
channels, the rate of harvest of the same or competing
products from the wild, and technological improvements.

In summary, the lack of biological information on
production technologies makes it difficult to determine the
most economically efficient input combinations for the
relevant range of outputs. In addition, more study is
needed on how to merge production information into proper
plant design, taking into account both biological and market
peculiarities.

Investment and Output

Obviously, output will be constrained by the amount of
capital available for investment. Business enterprises in
the United States usually require a 20-25 percent return
from new ventures. Returns from aquaculture generally are
not this high. Because the relevant technology is not well
advanced, industry may be further discouraged from investing
in aquaculture. Current investment may lock industry into
current technologies, whereas a new technology development
could greatly increase productivity.

In order to properly assess the profitability of
aquaculture, it is necessary to have accurate and timely
data on production processes and rates, types and costs of
inputs, expected improvements in technology in aquaculture
and other related food industries, and expected absolute and
relative price trends. Although much of this information is
not currently known, some is known but not publicly
available due to proprietary rights and some is scattered in
report files of various government agencies. If these data
were available, better assessments would be possible and
interested firms would be spared some pre-investment costs.
However, requiring disclosure of proprietary information may
not be justifiable and could well inhibit the development of
aquaculture.

Organizational Forms

Because many aquaculture processes are capital-intensive, plant size usually has to be large in order to achieve low costs per unit of product. For some species, large plants could increase output from 10-20 percent, thereby exerting considerable downward pressure on the price, and for some species, such increases would permit a complete takeover of the market. In some cases, large plants might have little impact on national markets but could still dominate marketing and distribution channels in particular sections of the country.

As already indicated, aquaculture for some species has achieved relative success. As these industries have developed, specialization has taken place. Such specialization is exemplified by the salmon industry, in which farmers in Washington produce young salmon to smolt size; these salmon are then purchased by salmon ranchers for grow-out, release, and recapture. The particular organizational form selected by participants may range from the small entrepreneur, for whom aquaculture is a "sideline," to a large corporation. Specialization and division of labor may result in the coexistence of many different sized firms as well as different organizational forms.

Labor and Management

Policymakers should be aware that technological advances may create unemployment in certain established industries and may necessitate relocation of workers. This does not mean that research into aquaculture production functions should be restricted, but rather that the full effects should be considered and planned for.

A production aquaculture facility needs academically trained personnel for only a small part of its staff. In some segments of the industry, job applications from formally trained persons far exceed the number of job openings because young people are enthusiastic about aquaculture. At the same time, many firms have difficulty finding personnel for routine production work, which involves on-the-job training.

Typical of most growing industries, the number of experienced managers is insufficient. However, the situation is particularly difficult in aquaculture in which it takes special training to develop expertise: a livestock producer does not automatically have the knowledge to produce catfish. Specialized training in health maintenance, water quality measurements, and production

techniques is necessary in order to grow fish. Equally
important, a prospective producer must be assured of a
market for the fish before any investments are undertaken
and this requires learning a whole new marketing system.

Aquacultural Sites

A basic misconception concerning aquacultural development in
the United States is that sites are plentiful. In most
cases, the sites that would be desirable or acceptable for
various aquacultural purposes are already used for many
activities and aquaculture is likely to be a marginal
competitor. Even in Hawaii, a state surrounded by ocean,
repeated surveys for sites have difficulty identifying
acceptable coastal areas where aquaculture represents the
best use. This is also the case for California and probably
will be found to be the rule, rather than the exception,
throughout the United States.

Where it might be possible for aquaculture to coexist
with other activities, the legal and institutional framework
is often inadequate for the establishment and protection of
aquacultural enterprises. Furthermore, tremendous
variations exist in state laws concerning such matters as
leasing, taxes, and harvesting. Overall, government
policies have a profound impact on aquacultural development
and production.

If aquacultural systems are to be land based (with water
pumped to the site), then profitability relies on factors
other than those that stem from the abundance of ocean space
and residual nutrients. Enterprises must compete for space
and investment funds with other investment opportunities in
the area.

Regulations Affecting Inputs

Aquaculturists and potential investors are seriously
concerned about regulations controlling chemicals and
pharmaceuticals. They feel their fish are vulnerable to
disease organisms that may develop resistance to the few
certified medications. They feel uncomfortable with a
situation where both governmental and private groups
routinely use materials not formally certified for
production use. Furthermore, they are apprehensive that the
few certified materials will be decertified, leaving growers
completely vulnerable to disease.

Because most aquaculture involves a product for human
consumption, many types of permits and clearances must be
obtained before actual operation can begin. Although such

permits are essential to human well-being, some interesting
cases can arise: raising shrimp in heated waste water from
nuclear power plants or culturing algae to feed oysters
using wastes of certain industrial agricultural processes
may be impossible with strict interpretation of current laws
even though both cases may be extremely safe if handled
properly.

A diverse list of regulations, restrictions, and permit
requirements often act as strong constraints to development
(see chapter 4). Aquaculture tends to be fragmented and
disorganized and dozens of regulatory bodies are involved
from the local to the federal level. The regulatory bodies
often do not coordinate with each other and at times their
demands conflict. Compliance with the requirements of many
regulatory bodies is costly. In addition, months and often
years of rigorous and expensive efforts must be spent to
obtain the necessary permits to initiate an aquacultural
production operation. Understandably, many potential
aquaculture operators and investors become discouraged and
give up their plans.

MARKETING

Problems of Market Development and Entry

The task of marketing is to move goods from production to
consumption at the time and in the form desired by
consumers. Problems naturally arise in new industries where
little information exists concerning consumer wants.
Temporary gluts or shortages often exist and irregularity in
supply prevents establishment of consumption patterns.
Creating demand for a new product may require skillful
advertising and market development campaigns. In general,
the introduction of a new product requires considerable
capital and regular supplies of uniformly high quality
products.

Because catfish farming is frequently cited as one of
the more successful areas of aquaculture, a summary of
marketing problems in that industry may suggest potential
problems for other fields of aquaculture. The general term
"unreliable market for catfish" covers a multitude of
complaints including fluctuation of prices over the season,
the difficulty in finding a buyer at harvest, and lack of
valid information regarding prices paid for catfish.

When catfish production was in the early stages, catfish
producers cited lack of buyers as the principal problem
involved in marketing fish. This problem often was not
recognized by producers until the fish were ready for
harvest. Producers were frequently so involved in

production that they failed to arrange in advance for sale of the fish. In spite of problems in marketing, many producers have not been interested in contracting for sale of their product, a practice that is prevalent in the poultry industry.

Catfish marketing stages include harvesting, transporting, processing, storing, and ultimate sale to the consumer. Producers are typically not concerned with market functions and prefer to leave marketing to the buyer.

Misunderstandings often result when processor harvest crews arrive and the pond has not been drained. Such lack of coordination is prevalent in producer-processor relations.

In the past, due to labor and transportation costs, processors would not harvest a pond that was more than 80 km from the plant, and they would only pick up producer-harvested fish from such locations if quantities exceeded 5 tons.

Processors purchase less than 50 percent of the catfish sold by commercial producers. The remaining commercially grown fish are sold to live handlers, restaurants, grocery stores, fish markets and individuals. In addition, substantial quantities of catfish are sold to fishermen who pay for the privilege to fish in ponds. An undetermined but substantial quantity of catfish is also grown for personal use. Market data are only available on sales to processing plants.

Grocery stores pay the highest average price for fish, and generally buy from wholesalers or processors. Commercial producers do, however, sometimes make direct sales to restaurants and individuals. Restaurants require a constant quantity over longer time periods and deal only with larger catfish producers or in some instances produce their own supply.

Processors have long desired a constant, year-round supply of fish. Until such a supply is available, market development will be somewhat restricted. In the future, processors may contract for and help finance additional tonnage. Now, catfish processors have become vertically integrated into production. Processors arrange production schedules from owned or leased ponds to fill slack periods of production by other growers.

Aquacultural producers may distribute products through traditional seafood markets or employ entirely different distribution networks. Further, as in the case of catfish, such markets may change over time. In the markets for more "traditional" seafood products, complaints are frequently voiced about the quality of the product reaching the final consumer. Because aquaculture affords the producer some degree of control over production, quality problems should be more amenable to correction.

If reaction to quality is negative, buyers and sellers of aquaculture products may choose to use different distribution networks or, in the extreme, to integrate vertically. In aquaculture industries, contractual arrangements involving specifications on product quality may be more prevalent than in more "traditional" seafood industries.

Distribution involves both the physical flow of product and the transfer of title to the product. As the product moves from producer to consumer, its form may change. If the product is moving into the restaurant market (where growth of fish sales seems to be occurring at a relatively rapid rate), the product or title to it may move through a different set of brokers and wholesalers than if it moves to retail outlets or overseas markets. Further, the form of the final product and the nature of the contractual arrangements may also be different. The industry (or firm) that is not aware of these differences or does not know the ultimate destination of the product may find its production inappropriate with respect to such characteristics as harvest size and flesh color.

In some seafood sectors and in some agricultural industries, grading standards have been established to deal with divergences between what is produced and what is "demanded," in terms of quality. The merits of such grading systems have not yet been evaluated and whether such systems will be applied to an expanding aquaculture industry is uncertain.

During the distribution process, the quality of aquacultural products often deteriorates because of poor storage and improper handling. Lack of information (especially by final consumers) on proper preparation of the product further detracts from the desirability of the product. Correcting these problems may be costly and the decision to improve storage and handling facilities will involve cost-benefit calculations. Although better storage facilities may improve the shelf life of an aquacultural product, the additional expenditure may not result in higher

profits to the industry. Nonetheless, it has been observed
that imported seafood products are sometimes preferred to
their domestic counterparts because of better handling
abroad.

Conduct of Firms and Market Structure

In some aquacultural industries there is intense rivalry
among producers. As new firms emerge and new techniques
develop, innovators are going to be reluctant to share their
discoveries (including market data) with competitors or with
independent (government and university) researchers. In
fact, some of the research proposed in this report may be
hampered by data limitations. Nonetheless, because public
policy affects aquaculture, the aquaculture industry may
benefit from sharing data so that public policy can be made
in an informed way.

However, government policy is developed in a broad
social context, and data on aquaculture will be only one of
the factors contributing to decisions affecting the
industry. As suggested earlier, some aquacultural endeavors
may be economically feasible only if conducted by large
firms. In examining the relevance of anti-trust laws to
such circumstances, the social costs and social benefits of
discouraging such market structures should be carefully
evaluated.

FINANCE

The financial community is traditionally conservative and
rarely can be induced to venture into areas where risks are
high, profits uncertain, and rules are not available to
guide decisions. This reluctance increases when management
ability has not been demonstrated, as is the case for many
aquaculture operations. Creditors are wary of lending money
to producers, such as aquaculturists, whose rights are not
securely protected by the relevant laws and institutions.
The result is that a potential aquaculture producer must be
in a secure capital position in order to obtain a loan.
Normal banking practice has required that working capital be
repaid in short term intervals such as 30, 60, or 90 days.
These periods bear no relationship to the realities of
producing crops that may have growing periods of one to five
years. Expansion capital is difficult to obtain where the
bulk of the applicant's assets are invested in underwater
lands with uncertain market value and in specialized
equipment with little value for another type of operation.

Obtaining Credit in the Private Sector

Economically, aquacultural operations in the United States can be divided into three groups. The first group consists of established profitable businesses that have a capital foundation and whose management skills qualify them for credit from lending institutions. The two prime examples in this group are successful oyster farmers in coastal states who enjoy laws that give them firm title to the underwater lands on which their crops are grown and whose rights to protect their crops from poachers are enforced and successful trout and catfish farmers who enjoy the same legal rights. Businesses in this group appear to have no difficulty in obtaining credit for working funds or capital for expansion although this has not always been the case.

Operations in the second group offer potentially high profits but also pose high risk. Such operations have been largely financed by lenders of risk capital or by large corporations that see sufficient merit in projects to qualify them for this type of investment. Several current trends suggest a decline in availability of financial resources. The before-tax rate of return that firms require to undertake new investments has been driven up by a number of forces. For example, greater risk premiums are required to compensate for bouts of macroeconomic instability (primarily wide variations in inflation and unemployment).

The third group consists of those projects with seemingly high potential that have not developed much beyond the laboratory stage. This group includes projects for artificial upwelling to produce abundant phytoplankton as food for shellfish and closed system aquaculture for oysters, clams, mussels, and various finfish such as yellow perch.

Projects in this group are primarily supported by research funds from the National Oceanic and Atmospheric Administration.

External Factors

Although the required rates of return on investment have risen, actual rates have declined. New environmental and safety requirements, rising prices, and uncertain supplies of purchased inputs, particularly energy, have escalated costs substantially.

Compounding the forces causing a greater shortage of funds are forces that channel funds toward bond markets, particularly special tax treatment that has become more attractive as inflation premiums have raised nominal

70

interest rates. Remaining funds for equity markets
increasingly tend to flow toward investments approved by
large institutional investors. Retirement and pension funds
are steadily accounting for a larger share of private
savings, and these funds are generally not invested in high-
risk ventures. Other factors inhibiting the flow of funds
to small high-risk ventures are state and federal
regulations intended to protect investors from fraud and
unexpected risk.

Under current conditions, the only significant growth in
the high-risk, high-profit segment of aquaculture will come
from large corporations, and even this will be limited
because large corporations investing in the field of
aquaculture appear to prefer to take over promising young
ventures rather than to initiate new aquacultural
development.

Public Financial Assistance

America's great success in agriculture is largely the result
of more than 100 years of research supported by billions of
dollars of public funds which has proved to be a very wise
investment. Aquaculture may merit similar treatment.
Ongoing research should be coordinated and directed to
ensure the most efficient use of funds. In addition, it is
extremely important that worthy projects that offer
substantial food contributions to society be supported
continuously until fully developed. Arbitrary time limits
have all too often resulted in projects being dropped before
they are completed.

No federal financial assistance programs include
aquaculturists specifically, but some programs available to
terrestrial farmers have been extended to include their
aquatic counterparts in a limited way. For example, some
states (such as Hawaii) have established loan or other
financial assistance programs for producers of aquacultural
commodities.

A commitment to encourage aquaculture development should
include consideration of the use of a specialized credit
system. The Farm and Rural Development Administration's
Insured Loan Program of USDA may be a useful model.
Experience with this program should be of great value in
designing an acceptable credit program.

An Example

A substantial number of small operations are based primarily
on research results that have not been previously applied

but that show theoretical profitability and offer an
opportunity for self-employment. An example of this
involves the European oyster, known as the flat oyster, and
highly regarded in France for beauty of shape and exquisite
taste. It is not exported to this country, but there is a
limited market at a high price (presently about twenty cents
per oyster to the grower) at restaurants that serve gourmet
food. At present, these oysters are being grown in the
state of Maine.

Maine has about four thousand miles of shoreline with a
number of rivers and streams bringing fresh water into many
estuaries sufficiently land-locked to prevent destructive
ocean wave action. Maine historically produced the American
oyster naturally, but oyster beds have been depleted for
many years, and no serious attempt has been made to revive
this crop. Maine does not have laws favoring private titles
to underwater bottoms and the present development with the
European oyster has come as a result of hatchery seed grown
at the University of Maine. More recently private
hatcheries in Maine and even California have also produced
seed.

Financing for these small operations has come
principally from private savings of the operators plus some
risk capital from individuals. The federal government makes
available to land farmers a variety of financial assistance
programs such as obligation guarantees, loans, and crop
insurance. Fairness dictates that such programs also be
made available to aquaculturists. If they are not,
aquaculture will be restricted to operators who can provide
financing from other enterprises.

Unlike traditional crop and livestock enterprises,
aquaculture has represented an almost "intangible product"
for lending agencies. Because fish are grown in water, it
is frequently difficult to ascertain the precise numbers and
weight of the product until harvest. Inventories can only
be estimated within a growing season. Further, due to the
relatively brief period of production, lending agencies have
not been able to establish a repayment history for the crop.

RECOMMENDATIONS

The lack of adequate marketing research and financial
programs, coupled with high costs of production inputs,
inhibit expansion of existing industry activity. Without
equitable public support, additional industry involvement in
aquaculture will be difficult to expect. Consequently, we
recommend:

* <u>Research to determine whether markets exist for species that are being considered for aquaculture. Such efforts should be concurrent with scientific research on those species.</u>

* <u>Federal support for public and private research on marketing aquacultural products..</u>

* <u>Research on the relationship between potential aquacultural products and the land, labor, capital, and operating inputs required to produce the species. This research should also examine the relationship between scale of operation and input requirements.</u>

* <u>Financial programs that are now legally available to terrestrial agriculture be equally available to those engaged in aquaculture.</u>

INTRODUCTION

Aquaculture in the United States has developed within a framework of laws and regulations designed to protect the rights of aquatic farmers in the same ways that the rights of terrestrial farmers are protected. These laws and regulations provide the opportunity to use owned or leased land, and the water thereon, to grow certain species of plant or animals under specified conditions.

Environmental laws are generally intended to protect waters from pollution and thereby provide a suitable environment for various uses, including culture of aquatic food species. The same laws ensure that effluents from aquacultural operations are controlled to maintain acceptable quality of public waters.

Health and safety laws protect the consumer of aquacultural products by requiring that fish and shellfish farms be located in areas where the waters are clean. They may further require that the products are processed and distributed under sanitary conditions and that working conditions in aquaculture production systems meet the same safety standards as in other industries.

Aquaculture is specifically authorized or controlled by many federal and state laws. Some of these laws and regulations may reduce the economic incentive for aquaculture by adding to costs, delaying operations, or by causing uncertainty. On the other hand, the development of aquaculture may also be constrained by the absence of laws. Some federal and state assistance programs available to terrestial farmers are not available to aquatic farmers because the applicable laws do not specifically include aquaculture or because administrators are unfamiliar with aquaculture. These programs range from government loans to technical advisory services.

Federal laws are applied differently in various geographical districts or regions. For example, in the

northwest, the U.S. Army Corps of Engineers currently
requires a permit for use of a hydraulic harvester for
clams. Such a permit is not required for use in the
Chesapeake Bay on the east coast.

More than twenty federal agencies or offices are
involved in activities directly or indirectly related to
aquaculture but until recently none had specific authority
to lead or coordinate federal efforts in the field.
However, the Food and Agriculture Act of 1977 (PL 95-113)
names USDA as the agency responsible for leading government
activities related to aquaculture.

Federal aquaculture laws generally fall into two
categories: those that apply to waters located beyond state
jurisdiction (e.g., the Fisheries Conservation and
Management Act of 1976) and those that authorize the conduct
of research and development programs and the establishment
of federal hatcheries. None of these federal laws or
regulations apply directly to private farming of fish,
shellfish or marine plants. However, other federal laws
related to land and water use, environmental protection,
health or safety affect private aquaculture ventures.

Trade and tariff laws and agreements may reduce the
economic incentive for development of aquaculture in the
United States by permitting entry of low-priced imports.
For example, imports of canned and frozen oysters from Japan
and more recently from Korea have increased from 50 mt in
1947 to 8,600 mt in 1973. Total domestic consumption has
remained at about 32,000 mt; increases in imports were
offset by decreases in domestic production.

Many of the laws and regulations that specifically
authorize, permit, control, or prevent aquaculture are at
the state level. State laws concerning aquaculture are
usually closely related to state authorities charged with
management of fisheries resources and hatchery programs are
therefore usually administered by state fish and game or
conservation agencies. In some cases, conflict exists
between the objectives of public resource management and
those of private aquaculture. Commercial fishermen or
recreational groups may oppose private aquaculture,
especially if private aquaculture requires allocation of
waters or submerged lands for private use.

Laws regarding aquaculture vary widely among states.
Oyster culture on the Pacific coast is entirely conducted on
privately owned or leased intertidal and subtidal lands. In
New York, Connecticut, Virginia, and Delaware, substantial
areas suitable for oyster culture have been allocated to
private oyster culture, but extensive public fishing waters
remain. Nearly all water bottoms suitable for oyster

culture in Maryland remain public fishing areas. North of
Cape Cod, commercial farmers cannot obtain sufficient
tidelands for production of oysters or clams. Private
salmon farming by the ocean ranching method has been
legalized in Oregon and California but not in Washington.
Ocean ranching is legal only on a non-profit basis in
Alaska.

In some states, laws designed for public fisheries
resource management have not been revised to take private
fish farming into account. For example, state statutes
specify that all fish existing in Louisiana are owned by the
state. This raises the question of ownership of catfish or
crayfish reared in private ponds, especially if they are
stolen or escape into public waters after a flood or storm.

In some cases, state laws written for conservation of
wild stocks prevent private culture, ownership, or marketing
of aquaculture species. Permits, licenses, and periodic
reports are required by state agencies for their
administration of laws specifically related to aquaculture
as well as laws related to land and water use, environmental
protection, health, and safety. In many states, more than
30 such requirements must be met before a producer may
legally commence operations.

The major aquaculture problems that arise from state
laws and regulations relate to the lack of uniformity of
laws in different states, the difficulty of obtaining
concise lists of legal requirements within a state, and the
difficulty of obtaining the many permits and licenses
required for aquaculture ventures. A recent decision by the
U.S. Supreme Court strikes down state laws that limit
licensing to state residents, thereby removing a potential
barrier to development of aquaculture by non-residents.

The following section describes relevant laws in general
terms and provides the basis for identifying those that may
inhibit the development of aquaculture.

COASTAL LAND AND WATER USE

Legal constraints on coastal land and water use for
aquaculture are many and diverse and arise from several
basic sources. A comprehensive study of them is not
feasible here but the general sources and nature of these
constraints can be described.

Legal constraints arise primarily from the existence of
conflicting uses. Regulation of several individual uses
provides the legal structure within which new uses must fit.
Thus, multiple uses mean multiple regulation of water and

land areas, often to the detriment of uses not anticipated
or provided for when rules or regulations were developed.
Compounding the problem is the fact that diverse uses are
frequently regulated on several levels (international,
national, state, or local) at one time without adequate
coordination.

The laws and regulations established to regulate
conflicting uses give rise to three basic classes of legal
considerations: (1) legal problems related to the right of
access; (2) legal problems related to property rights; and
(3) other governmental controls affecting management and
use. These are discussed below in conjunction with land,
water, and submerged land use and their relation to
aquaculture. In examining these issues, consideration is
given to the various sources of legal constraints, such as
international, federal state and local law.

Upland Use Problems

Land access rights are generally controlled by state or
local law. However, these rights are also affected by some
federal enactments in the form of federal incentives or
guidelines, such as the Coastal Zone Management Act (CZMA
[16 USC 1451 et seq.]), or legislation that may be aimed at
curbing pollution, such as the Federal Water Pollution
Control Act (FWPCA), the Ocean Dumping Act (ODA), and the
National Environmental Policy Act (NEPA).

Zoning is one of the most significant of state police
powers although it is commonly delegated to county or
municipal governing bodies. All privately owned land is
held subject to existing and future zoning (_Village of
Euclid_ v. _Ambler Realty Co_ [275 US 365]). Zoning
ordinances, based on the police power, must conform to
police power standards. A zoning ordinance will thus be
invalid if it is unreasonable, arbitrary, discriminatory, or
confiscatory (_Pennsylvania Coal_ v. _Mahon_ 1922 [260 US 393]).
Absent these features, a zoning ordinance is valid to
control the use of private land. Conceivably, a landowner
might be able to obtain a variance if the land cannot yield
a reasonable return as zoned, or the ordinance may be shown
to be otherwise unreasonable and the use claimed would not
alter the essential character of the locality. Zoning
designations should be, and often are, made under the
guidelines of a local master plan. Such plans, in turn, are
often influenced either by aesthetic values held by a
community with a view to human habitation or by the need to
broaden the tax base. Thought is seldom given to potential
new uses (such as aquaculture) and some difficulty may be
encountered in fitting new uses into existing plans and
categories of use.

77

Coastal Zone Management Act (CZMA)

Zoning, as it affects aquaculture, is directly influenced by
CZMA. That Act affects the use of the coastal zone, which
is defined as the coastal waters and the adjacent shorelands
strongly influenced by each other and in proximity to the
shorelines of the several coastal states (16 USC 1453).

The Act provides federal incentives for states to
establish the appropriate machinery for land and water use
planning in the coastal zone. No specific method is
required, but to qualify for federal help, the state must be
prepared to draw up a specific zone use plan, including an
inventory and an identification of the means it will use to
control land and water. Furthermore, the state must
demonstrate that it has authority through an agency or
agencies, including local governments, to administer land
and water use regulations, to control development, and to
resolve conflicts among competing uses. It must also have
the power to acquire fee simple and lesser property
interests through condemnation or other means (16 USC 1455).

The state may establish criteria and standards for
local implementation (with state review and enforcement) or
it can directly assume state land and water planning
regulation. It can also establish procedures for state
administrative review for consistency of all development
plans proposed by state or local authorities, or by a
private developer, with power to approve or disapprove. The
federal lever under CZMA is funding. A state qualifies for
planning and administrative funds if it complies with the
guidelines set forth in the Act.

From the standpoint of aquaculture, the Act is
advantageous in that it encourages rational, shared use in
the coastal zone and flexibility with regard to changing
conditions and new uses. However, land planning mechanisms
generally favor public use of land or private use that will
broaden the tax base. Although CZMA efforts have tended to
disregard these considerations and have favored water-
dependent uses, coastal lands are more likely to be
designated for habitation, public, industrial, or
recreational purposes than for purposes such as aquaculture.

State Land Laws

State land management legislation may also affect land use
through zoning. Florida, for example, has legislation (Fla.
Stat. 380.012-380.12, 1972 [Florida Environmental Land and
Water Management Act]) that gives the state power to elevate
land use decisions to a higher level if the limited area of
land is a subject of "critical concern" based upon

significant impact upon environmental, historical, natural,
or archeological resources of importance or is of otherwise
broad state concern (Fla. Stat. 380-05). Thus the use of
land may now be subjected to scrutiny on the local, state,
and federal levels. Use of land for aquaculture purposes
may be precluded as master schemes, now under development in
many states, are completed.

Common Law

In addition to zoning ordinances, certain common law
restrictions can apply to land. Some of these might arise
from doctrines prohibiting use that could be construed as a
public or private nuisance[1] or from restrictions voluntarily
placed upon land at a prior time, such as easements,[2]
profits,[3] covenants running with the land,[4] or equitable
servitudes.[5] Such restrictions could prevent or limit
otherwise suitable land from being used for aquaculture.

Access

Access to water from land along coasts, rivers, and lakes is
rarely a problem. In fact, riparian rights (property rights
that apply to the use of water) in most states assure the
littoral owner access to and from the water by means of
improvements such as piers and docks and dredging of
channels to land (United States v. Rands 1967 [389 US 121],
City of Philadelphia v. Standard Oil Co. of Pa. 1935) (see
also Appendix A). The riparian owner is also assured rights
of navigation and fishing, preference in the development of
submerged land, and freedom from interference by neighboring
owners. The owner's use of the land could, however, be
hampered to some degree if the land is subject to access
through it by the public. Under the law of easements, the
public may acquire, by use over many years, rights in
private property along the shore. Dedication to the public
that results from nonprohibited use over a period of years
is a broad principle that is also applicable. Furthermore,
legislation could be (and has been) used to declare public
rights of access to beach areas (Fla. Stat. 375.031).
Finally, common law limits private ownership of land at the
high water mark, thereby guaranteeing public access along
beaches between high- and low-water marks.

Intertidal and Submerged Land Use

Limitations on the use of submerged lands can significantly
affect fish farming. Some enterprises, such as oyster and
sponge culture, require use of such lands; others, such as

finfish culture, use submerged lands only if special
installations or ponds must be constructed.

International Law

The Continental Shelf Convention of 1958 (CSC [499 UNTS 311,
15 UST 471 TIAS No. 5578]) assigns resource rights,
including rights with respect to sedentary species, to the
coastal nation seaward to a certain limit. Rights to fish
resources extend 200 miles seaward of the coastline.
Nations are free to regulate resources within these limits.
The coastal nation has the exclusive authority to explore
and exploit the living and nonliving resources of the
adjacent continental shelf. This authority has been
exercised in the United States primarily with respect to oil
and gas, but there is also federal legislation that
restricts taking sedentary species.

Federal and State Law

Both the federal and state governments have rules regarding
any activity affecting navigable waters. For example,
federal approval--through the U.S. Army Corps of Engineers--
must be obtained prior to the construction of a dam or dike
or for any supporting piers or wharves within the navigable
waters of the United States (33 USC 401.403). An
aquaculturist who wanted to dam, dike, screen off, stake, or
otherwise exclusively use navigable waterways by building
obstructions would also have to obtain the permission of the
state government. State approval, however, in no way
supplants the requirement for approval by the Corps of
Engineers (Cummings v. City of Chicago 1903 [188 US 410]).
In addition, a riparian landowner who fences off an area in
such a way that it obstructs the free passage of fish up or
down a stream may be subject to a common law action and
required to remove the obstruction (State v. Haskell 1911
[79 Atl. 852]).

     Grouped together, the Outer Continental Shelf Lands Act
and the Submerged Lands Act (43 USC 1301-1325) divide
leasing and regulatory responsibilities for submerged lands
between the state and federal governments. Under the
Submerged Lands Act, states have the power to lease areas
within their jurisdiction, generally three miles seaward
from the coastline. In states (like Florida) where leases
can be granted for purposes of aquaculture, leasing
conflicts may arise and competition for areas may exist.
Furthermore, the question must arise whether there can be
multiple use leasing of the same area (Louisiana has had
some experience in dealing with the problem of mineral
leases as opposed to oysters leases). As more and more

states begin to regulate or provide protection for fish culture, clarification of this issue will be required.

Water Use Problems

If an aquaculture operation involves the use of water areas not including the bottom (for example, by use of pens or rafts), a different set of legal constraints apply. The constraints have application in dealing with potential conflict between aquaculture and such other uses as navigation, fishing, and recreation.

International Waters

It is unlikely that substantial "contained" fish farming would be conducted beyond the limits of national jurisdiction in the near future. However, if that should be the case, the customary and conventional law of the sea would apply: in particular, the High Seas Convention of 1958 (450 UNTS 82, 13 UST 2312, TIAS No. 5200 [1962]). This convention, among other things, provides guarantees for freedom of navigation and freedom of fishing. However, the listed freedoms are not the only ones that could be exercised, and each must be exercised with reasonable regard to one another. Aquaculture may require the use of an area of water to the exclusion of navigation or fish capture, but that fact alone will not render the use contrary to international law. The question is clearly one of reasonableness, taking into account such factors as the size of the area, its location, and the duration of use. At the present time, the same problem exists with respect to oil and gas extraction, and reasonable limits to navigation on the high seas have been voluntarily complied with without contention.

The high seas may thus be used for fish rearing, but the operator would be subject in some respects to the traditional tort law of admiralty for damages the operation may cause to vessels or persons in the area.

Although the general principle of freedom of fishing applies on the high seas, agreements with foreign nations under the Fisheries Conservation and Management Act of 1976 (FMCA [U.S. Congress, House 1976]), and the regulations and management plans generated thereunder, may have an impact upon high seas fish farming operations. Regulations under this Act apply to waters that lie more than three, but less than 200 miles from shore. These regulations, through their control of fishing, could affect aquaculture in two ways: if excessive catches are permitted, the nature of the biomass may be changed or the general level of productivity may be reduced; if gear restrictions are not compatible with capture operations, aquaculture in particular areas could become impossible.

Public or private ocean ranching may be affected by international laws or agreements. For example, during their migrations, salmon may feed beyond the 200 mile fisheries zone authorized by the FCMA, and they may also migrate northward into the coastal zone of Canada. During those migrations, salmon are vulnerable to capture by fishing fleets of many nations. Clearly, the international implications of salmon ranching need close scrutiny.

Finally, customary international law recognizes the right of a nation to close certain areas of the high seas for limited purposes, thus removing the availability for fishing. Examples are areas necessary for national security, weapons testing, and the like.

Federal and State Waters

International law protects navigational interest of other nations within the territorial limits of the United States. Under the Geneva Convention on the Territorial Seas and Contiguous Zones, a coastal state is under an obligation not to hamper innocent passage. But a state may enact reasonable laws regulating the usage of such waters, so long as the cumulative effect does not amount to hampering passage.

For U.S. citizens, federal and state law guarantees public rights to use navigable waters of a state for navigation (_Merrill-Stevens_ v. _Durken_ 1912 [57 So. 428]). The question is thus raised whether this right is superior to any others that might be exercised within state limits. The right of navigation, while held in common with all citizens, is also subject to reasonable exercise of state police powers, and the right may thus be limited in favor of the general welfare (_State_ v. _T.O.L. Inc._ 1968 [206 So. 2d 69]). The difficulty is that in most states there is at present no adequate mechanism for deciding priorities.

The public also has a common right to fish in navigable state waters (_Anderson_ v. _Reames_ 1942 [161 SW 2d]). Like the right of navigation, however, that right can be limited by placing reasonable restrictions on taking fish, including licensing and conservation measures. Restriction of fishing is clearly within state police powers, but most state laws are silent on the question of the relative positions of capture and culture operations and provide the fish farmer with little or no protection. Florida is unusual in this respect, providing by statute for the leasing of state

submerged lands and water column specifically for use in
aquaculture (Fla. Stat. 253.67-253.75). The law has not,
however, been used and remains untested. Fishing
regulations with respect to areas, closures, spawning
grounds, gear, health restrictions, and methods of capture
also limit aquaculture in many states.

The FCMA, although not generally applicable to fishing
in state waters (waters within three miles of the shore),
may be so applied if a fishery is primarily conducted
outside of state waters and if the state is failing to
implement plans developed under the Act. The FCMA did
establish regional fisheries management councils charged
with developing specific plans that will result in quotas
and ceilings of fish catches and recommendations for their
allocation.

Recreation

Laws providing for the acquisition or setting aside of land
and water for recreation or use as marine preserves clearly
will affect the future of aquaculture. Public recreation,
like fishing and navigation, is subject to state statutes
and court decisions, and a state could restrict recreational
use in favor of some other public interest. Whether that
doctrine would be applied in favor of private enterprise,
however, is somewhat doubtful. In the future, increased
pressure for recreational use of water may result in laws
that will set aside increasingly large areas for restricted
use, thereby directly limiting aquaculture development.
Development may be further limited by water pollution or
destruction of the detrital food web that can result from
these other uses.

WASTEWATER EFFLUENT CONTROLS

Aquaculture creates a significant source of pollution. High
concentrations of fish and shellfish grown at rapid rates
create concentrated and increased waste load. Unless wastes
are quickly and efficiently removed from aquaculture
systems, disease or oxygen depletion may result.

Waste products from aquaculture operations can come from
ponds or raceway cultures. Wastewater from ponds may
contain unused food, animal waste products, and chemicals;
it is often disposed of in bodies of water near the
aquaculture operation. In raceway cultures, the continuous
flow of water carries by-products of the operation away in
the stream flow.

State and federal laws control this source of water
pollution. The costs of developing pollution abatement
technology to meet the requirements of these laws are
considered prohibitive by the aquaculture industry. The
industry has urged the U.S. Environmental Protection Agency
(EPA) and state agencies to reduce water pollution standards
for its particular operation, or exempt the operations
altogether, until the appropriate pollution control
technology is developed. Some segments of the industry have
also urged special treatment because of the small size of
many aquaculture operations, the infrequency of discharges,
and the relative newness of the industry.

Federal Water Pollution Control Act

The Federal Water Pollution Control Act Amendments of 1972
(FWPCA [33 USC 1251]) and its implementing regulations (see
40 CFR), are administered by EPA and certain cooperating
state agencies. Any discharge of a pollutant from a point
source into U.S. waters is prohibited unless made pursuant
to a National Pollutant Discharge Elimination System (NPDES)
permit from EPA or from a state agency that has been
delegated permit program authority by EPA. In almost all
cases, effluents from ponds and raceway cultures are covered
by the FWPCA. Pens or rafts in open water are not now
considered to be point sources requiring a permit, although
they may be subject to state law and regulation. They may
also be subject to regulation as non-point sources of
pollution.

Effluent limitations set forth in point-source discharge
permits are based on EPA effluent limitation guidelines and
standards. These regulations limit both the type and
quantity of discharge. Effluent limitations have been
published for over 20 industrial and user categories, many
of which are divided into numerous subcategories. These
limitations determine the kind of pollution control
equipment or procedures that must be used by a facility,
which may result in specific costs to industry.

EPA has not as yet proposed effluent limitation
guidelines and standards for aquatic animal production
facilities. Effluents from the processing of aquatic
animals are covered by EPA regulations, but effluents
resulting from growing the animals are not.

In the absence of specific effluent limitations for an
industrial category, each EPA regional administrator or the
permitting state's director (subject to the EPA regional
administrator's veto) must determine the appropriate
effluent limitation for each permit application. Under
these circumstances, differences between state standards and

EPA regional administrators' judgments create conflict. For
example, in the state of Washington, the Department of
Ecology wants to reduce the effluent standards for certain
hatcheries, but the EPA regional administrator will not
allow this, and a law suit is in progress.

According to EPA authorities, a revised development
document for effluent limitations pertaining to fish
hatcheries and farms is currently planned. A revised
environmental impact study will also be prepared. These are
not proposed regulations, but technical information
documents.

Almost all aquaculture operations will require an NPDES
permit although, in 1976, EPA redefined aquatic animal
facility to exclude certain small operators. Until effluent
limitations are established, the specific suggestions in the
development documents will probably be followed by most EPA
regions. Within a Region, states authorized to issue
federal permits may have different effluent standards,
although many state programs had not adopted special
aquaculture controls prior to the passage of the 1972 FWPCA
amendments and these states will now rely upon EPA guidance.

Aquaculture Exception

The FWPCA specifically permits discharge of effluents into a
defined area of the navigable waters from a "managed
aquaculture project." These discharges may exceed effluent
standards because of the benefit to society from increased
plant and animal production. The "project" must be under
state or federal supervision and have received a permit
pursuant to the regulations. Pollution discharges must
occur in a defined project area and not contribute to water
pollution outside the designated area.

This provision was motivated by experiments with the use
of heated effluents from power plants and waste discharges
from sewage treatment plants to assist in growing aquatic
organisms. It could be applied to an aquaculture operation
where fish-processing plant water could be used to assist
growing a new crop of animals in a defined (or designated)
area.

Effects of FWPCA on State Powers

Because EPA's program under FWPCA is mandatory and
comprehensive and offers states a positive role in the
federal program (states can issue the federal permits if
they have adequate laws and standards), state water
pollution control efforts are, for the most part, an adjunct

to the federal program. However, some remaining state
powers are worth mentioning. State water quality criteria
can be used in making specific permit decisions, and states
can review activities of applicants for federal licenses and
permits to be sure state water quality criteria and
standards are not abridged. For non-point sources of
pollution, state or local level control plans must be used
with planning and technical assistance from EPA. Further
public health laws are administered at the state and local
levels and agencies can order the abatement of a polluting
activity if human health is endangered in any way.

PUBLIC HEALTH AND SAFETY

Health and Sanitation Requirements

A number of federal and state laws applicable to aquaculture
operations or products have been enacted to protect the
health and safety of consumers or aquatic farmers.

Federal laws, including the Federal Food, Drug and
Cosmetic Act (FDCA) (21 USC 346a) administered by the Food
and Drug Administration (FDA), concern the quality and
safety attributes of various food commodities including
those produced in aquaculture systems. All fish or
shellfish farmers are responsible for assuring that products
produced in aquaculture systems are pure and wholesome to
eat and produced under sanitary conditions.

According to FDCA, a food is illegal if it contains a
natural or added deleterious substance that may be injurious
to health or unsafe. Food additives must be cleared for
safety before they may be used in a food or become a part of
a food as a result of the processing, packaging,
transporting, or holding of food. Excessive residues of
pesticides, heavy metals, or other toxic chemicals are
prohibited. Animal feeds that contain drugs to control
disease must meet the same regulatory requirements whether
they are used for aquaculture operations or meat and poultry
husbandry.

All drugs used on aquatic food species must be safe for
the consumer and safe and effective for the aquatic species.
Drugs that are considered to be new animal drugs, as defined
in the Act, must be approved by FDA before they can be
marketed. Animal drugs, however, may be used for
investigational purpose after they have been properly
cleared by FDA's Bureau of Veterinary Medicine. The role of
natural marine chemical compounds merits attention.

Aquacultural products can have problems meeting public health standards. Major obstacles arise from biological or chemical contamination of the water, diseases in the fish, and pharmaceutical residues from commercial feeds or water additives.

The degree of potential consumer hazard for these categories varies according to the type of aquaculture practiced. For example, water quality would be the most important category for oysters cultured in tanks, ponds or estuaries, but pharmaceutical residues may be of greater concern when fish are grown in closed systems requiring feeds containing pharmaceutical agents. Diseased animals may not necessarily be a cause for public health concern, but sale of such products may be prohibited under the Act if the edible flesh of the animal is adversely affected by the disease.

Fish and fish products may be unsafe or unwholesome due to excessive residues of pesticides, chemicals, or poisons resulting from ingestion, treatment with drugs, or exposure to pollution. Some of the currently identified contaminants in fishery products include heavy metals, pesticides, microorganisms, petroleum compounds, radio-active materials, sewage, and industrial chemicals such as those from pulp mills (see also Chapter 2).

Fish have been described as biological sponges. They can absorb many inorganic and organic materials through their gills as well as from the intestinal tract and some absorption probably also occurs through the skin. Fish directly reflect the character of the environment from which they originate and represent various risks to consumers.

Production of marine or freshwater food commodities in aquaculture systems may pose specific problems for the fish farmer that do not exist for the fisherman harvesting under natural conditions, but the product reaching the consumer should meet the same standards of quality and safety as products that are harvested under natural conditions.

Public health laws relate to consumer health but provisions and administration of these laws may affect producers in several ways such as: (1) approval of areas for fish or shellfish culture; (2) approval of water supplies for use in processing plants; (3) design of processing facilities; and (4) operation of processing plants.

Site Approval

Aquaculture operations, especially those related to
molluscan shellfish production, can legally be conducted
only in areas approved for this purpose by a state agency
acting under federal guidelines. Site approval is an
important consideration in oyster, clam, or mussel
production. Culture of crustaceans and finfish are not
included in the shellfish sanitation program and
requirements for growing these species are less stringent.
Nevertheless, where water is contaminated by pesticides,
chemicals, or domestic wastes, production and harvesting can
be prevented by public health agencies. In the James River
in Virginia, harvesting was prohibited by the state health
agency because of the presence of the chemical Kepone.
These prohibitions would have applied equally to
aquacultural products as to the harvest of wild stocks.

In the case of molluscan shellfish, elaborate procedures
have been developed in a cooperative state-federal-industry
program for certification of uncontaminated areas. Such
certification has considerable significance for the rearing
and harvesting of oysters, clams, and mussels.

One part of FDCA, the Delaney Clause, generally states
that no substances carcinogenic to humans or test animals
shall be added to food products. Because radionuclides are
known to be carcinogenic, this law would appear to inhibit
the use of waste heat from nuclear energy plants for
aquaculture and thereby may preclude the use of promising
aquaculture areas.

The siting and operation of aquaculture operations may
also be affected by the availability of approved water
supplies for use in processing plants.

Processing Plant Requirements

State health agencies and the FDA specify requirements for
the design and construction of processing plants. These
requirements usually include the use of stainless steel or
other smooth surfaces that can be readily cleaned after a
day's operation. Other requirements include the
availability of steam and chlorination. Requirements have
become more strict in recent years and many small oyster and
clam shucking houses have closed because they lacked the
funds to buy the required equipment or processes.

State and federal health agencies also specify
operational procedures to be followed in processing plants
to assure clean and safe products. All of these
requirements add to the cost of the product and in some

cases may adversely affect the profitability of aquaculture
operations.

Use of Drugs and Chemicals

Aquaculture is constrained by the lack of an adequate
battery of effective and certified chemicals and
pharmaceuticals. Chemicals and pharmaceuticals come under
FDA jurisdiction and must pass rigid and highly specific
certification requirements. Only a few chemicals and
pharmaceuticals have been certified by FDA (and in some
cases EPA) for commercial aquaculture use, and some agents
that have been certified require recertification in the near
future. Many others that are not certified are used
illegally.

Certification is an expensive and time-consuming process
and manufacturers of chemicals and pharmaceuticals for
aquaculture are reluctant to initiate certification
procedures when the immediate market is relatively small.
If separate clearance procedures were established for minor-
use compounds, production of such compounds might be
increased.

FEDERAL ACTIVITIES AND SERVICES

More than 20 federal agency offices or organizations are
involved in activities directly or indirectly related to
aquaculture. The most relevant agencies are within the
Department of Commerce, the Department of the Interior, and
the Department of Agriculture. At the federal level,
interagency coordination and communication has been poor. A
first attempt to establish regular formal communication and
cooperation among federal agencies is the Subcommittee on
Aquaculture operating within the Committee on Atmosphere and
Oceans (CAO), part of the Federal Council on Science and
Technology. The Subcommittee, chaired by NOAA, has served
primarily as a communications forum for interested agencies.

Advisory Services

The Extension Service (ES) of USDA accounts in part for
current U.S. preeminence in world agriculture today.
Establishment of the Sea Grant Program in 1966 brought about
creation of a parallel Marine Advisory Service (MAS) in
NOAA. Currently, aquaculture extension authorities are
divided among agencies.

Freshwater or inland farming is under the jurisdiction
of USDA's ES, but existing extension offices are often

neither prepared nor inclined to provide services to
aquaculture. Some extension services are provided by the
U.S. Fish and Wildlife Service of the Department of the
Interior.

Marine and coastal zone aquaculture is under the
jurisdiction of the National Oceanic and Atmospheric
Administration in the Department of Commerce. Aquaculture
in the Great Lakes is also under the jurisdiction of the
Department of Commerce because, under the terms of the
Marine Research and Engineering Act of 1966 (PL 89-454),
Congress classified the Great Lakes as oceans.

The ES (involving about 19,000 persons) is oriented to
the terrestrial animal or plant farmers. The MAS is
oriented to fishermen. Aquaculturist, in theory, should
benefit from both of the services; in practice, they tend to
be ignored.

Clearly, more thorough, complete and sophisticated
technological assistance should be available to the
aquaculturist.

Permit Procedures and Requirements

The procedures required to obtain permits and licenses have
been a severe deterrent to aquaculture. Because aquaculture
operations deal with food production, water supply, the use
of navigable waters, and effluent discharge, they are
regulated by agencies concerned with wholesomeness of food,
public health, water purity, land use, and pollution
control. As we have discussed, these regulations sometimes
occur at all three levels of government--federal, state, and
local. Overlap occurs because each level of government
feels it has an interest in the resource or health question.
Many environmental, health, and consumer-oriented programs
are new and are still defining areas of responsibility and
coordinating with other agencies. Further, the rules of the
game are being changed constantly as Congress, state
legislatures, and courts respond to pressures of many
diverse users and protection organizations.

Aquaculturists and other industrialists have claimed
that permit problems of redundancy, uncertainty, and
inflexibility have severely affected the industry.
Government has been sensitive to these claims. In the past
few years a number of government agencies have implemented
programs designed to simplify procedures to address the
problem.

General Permits

Both the U.S. Corps of Engineers and EPA have initiated
programs where "general" permits are issued for a class of
user in a particular area. EPA's proposed regulations
(published on February 4, 1977) propose a five-year general
permit to be issued for point sources of pollution in the
"separate storm sewer" and "agricultural activities"
categories. Once EPA has issued such a general permit,
members of the user group need not obtain separate permits
within the five-year period.

Although permits cover construction that has major
environmental impact, the Corps' general permit program
covers generally practiced activities that cause only
minimal adverse environmental impact. The Corps has to
issue a permit for any construction that might affect
navigable waters and general permits have been issued in the
Great Lakes region to cover small groins emplaced in waters
for erosion control by private owners. In Louisiana, the
Corps is considering general permits for cattlewalks, small
slips and piers and various aspects of oil and gas
development operations in the marshes.

Consideration could be given by EPA to the issuance of a
general permit for certain subcategories of aquaculture
operations where the impact on the industry of requiring
separate NPDES permits is not sufficiently offset by the
gains in better water quality.

Central Coordinating Agency

At both the state and federal levels, special new agencies
or new units within existing agencies have been formed for
the purpose of coordinating environmental and resource
permit reviews. At the state level, a division of the
governor's executive office, or a branch of a state natural
resources or environmental protection agency, usually serves
the function of a clearinghouse to inform users of
environmental permits required from state and local units
and usually circulates permit applications to the
appropriate offices. In the state of Washington, there is
an Environmental Coordination Procedures Act, administered
by the state's Department of Ecology: applicants are told
which permits they need, who they must see, the technical
information required and whether a state environmental
impact review is necessary.

In New Jersey, within the state's Department of
Environmental Protection, a Division of Environmental
Coordination informs users of the type of review the

proposed project must undergo and the information needed in
an environmental assessment. The requirement for a separate
state environmental assessment can be waived if a federal
environmental impact statement must be prepared.

Texas is proposing an interesting coordination device as
part of its coastal management program. An "activity
assessment routine" would be established by statute to be
administered by an interagency natural resources council.
This routine would establish the type of technical analysis
that must be prepared for a proposed project and would list
relevant requirements that have been established by state
resource agencies with jurisdiction over the activity.
Under this system one assessment could meet the requirements
established by all the relevant state agencies.

A similar central coordination function was established
in the Federal Deepwater Port Act of 1975. The Coast Guard
is the lead agency for granting a license to develop a
deepwater port beyond U.S. territorial waters; it is
mandated to solicit and receive the views of all other
agencies with respect to applications for a deepwater port
and to prepare the environmental impact statement.

With respect to aquaculture, state fishery agencies and
university extension services have performed an advisory
function for users regarding environmental, health, and
consumer laws. Coordination and advice regarding the permit
system might best come from specialists who know the
industry operations and economics, rather than from the
regulating agency. Any new efforts in aquaculture should
stress advisory services. The capabilities of advisory
agents to deal with pollution issues should be enhanced.

The Lead Agency

On September 29, 1977, the President signed into law the
Food and Agriculture Act of 1977 (PL 95-113) which assigns
federal lead-agency responsibility for aquaculture to the
U.S. Department of Agriculture. This action is a landmark
position with its clear-cut assignment of federal
responsibility to a single federal department and should do
much to eliminate the confusion of purpose, the diffusion of
energy, and the inequity under the law which together have
stultified the growth of U.S. aquaculture activities.
Whether or not that agency is, or remains, the Department of
Agriculture does not alter the thrust of our
recommendations. We encourage the lead agency to assume its
responsibility and promulgate a fully coordinated federal
aquaculture program.

Such a program should seek relief from inequities under the law for aquaculture, and seek to overcome scientific, technical, economic, legal and administrative constraints identified in this report.

Our specific recommendations for actions and programs to be undertaken by the lead agency follow this section. We believe those actions and programs must be carried out within a broad framework of developing the aquacultural industry. The Committee believes that within this framework, a lead agency should:

•   Develop, direct, and coordinate all U.S. aquaculture activities within the federal government;

•   Represent and report on executive branch programs in this area to the President and Congress on a regular basis;

•   Represent the United States in international matters relating to aquaculture;

•   Recognize and enhance aquaculture programs in other relevant federal agencies;

•   Coordinate efforts directed at obtaining funds from Congress for aquaculture activities;

•   Lead efforts at all levels of government to ensure that aquaculture interests are adequately considered in the development of regulatory requirements for protection of the environment, public health, land use management, etc.;

•   Ensure adequate long-range research and development in aquaculture which requires a level of funding that cannot be sustained by private business; and

•   Ensure that aquaculture enjoys parity of opportunity with related food production systems in terms of regulations, governmental support and insurance programs, and small business development programs.

RECOMMENDATIONS

In previous chapters, we have noted that aquaculture is constrained by technical, scientific, and economic factors and have made recommendations to overcome some of these constraints. We have not canvassed all possibilities, but believe that the following actions are sound. We recommend:

•   The lead agency take steps to develop
appropriate means to coordinate activities among
other federal agencies with aquaculture programs.

•   The lead agency annually submit a report on
aquaculture in the United States to the President
and the Congress containing a review of existing
problems, progress to date, the allocation of
funds, and an outline for action.

•   The lead agency develop an integrated program
that recognizes the assets and skills of existing
aquaculture programs within other federal agencies.
Such a program should include a National
Aquaculture Plan that specifically considers the
resources of other federal agencies and contains a
timetable for action.

•   The lead agency disseminate information
concerning federal and state permits and licenses
necessary for aquaculture operations and facilitate
their acquisition.

•   The lead agency develop and promulgate
guidelines for state use in enactment of laws and
regulations relevant to aquaculture development.

•   The lead agency encourage the U.S.
Environmental Protection Agency to review and
establish specific effluent guidelines and
standards for aquatic animal production.

•   The lead agency initiate with the U.S. Fish
and Wildlife Service a review of the Lacey Act (18
USC 42) to: (1) identify species of importance to
aquaculture; (2) establish a mechanism permitting
entry of such species if the public interest will
be best served by such entry; (3) develop and
distribute to state authorities model regulations
for interstate aquatic species disease
certifications, and (4) compose and publish a
listing of species whose entry into the United
States is controlled by statute, but which are
approved for use in aquaculture under specified
conditions.

•   The lead agency take legal and administrative
steps to clarify the rights of both public and
private hatcheries to harvest the ocean stocks they
have planted.

•   The National Oceanic and Atmospheric
Administration, through its office of Coastal Zone

Management, issue guidelines to states for
developing Coastal Zone Management plans that
recognize the potential of aquaculture.

•   The aquaculture activities of the National
Oceanic and Atmospheric Administration's Sea Grant
program focus on the long-term research and
development requirements of aquaculture, rather
than on the short-term (three year) problem solving
program currently supported.

•   The Agricultural Extension Service of the U.S.
Department of Agriculture be expanded to include
services for aquaculturists.

NOTES

1. Nuisance is a field of tort liability. A private nuisance can arise out of an intentional invasion of one's interest, or a negligent one, or conduct which is inherently dangerous. The essence of a private nuisance is an interference with the use and enjoyment of land. Thus if one uses one's land in a way to cause such an interference, such as by vibration, blasting, flooding, or causing pollution, etc., one may be liable to one's neighbor. A public nuisance is a nuisance affecting the general public, as in instances involving public health or safety, such as the maintenance of a malarial pond, or the storage of explosives. See, generally, Prosser, <u>Law of Torts</u>, Ch. 17.

2. "An easement is an interest in land in the possession of another which: (a) entitles the owner of such interest to a limited use or enjoyment of the land in which the interest exists; (b) entitles him to protection against third persons from interference in such use and enjoyment; (c) is not subject to the will of the possessor of the land; (d) is not a normal incident of the possession of any land possessed by the owner of the interest, and (e) is capable of reaction by a conveyance." (Restatement, Property, S450.) In more general terms, an easement is a property right to use the land of another for a special purpose not inconsistent with the general property right in the owner of the land: in the vernacular, a "right of way."

3. A profit (or profit a prendre), is an easement "plus." It is the right to use the land of another by removing soil or its products such as gravel, mineral, and timber.

4. A covenant that "runs with the land" is a restriction placed in the deed to property that subsequently binds purchasers of that land. It can provide a benefit or a burden to such purchasers.

The subject is highly technical and difficult. For
a complete, but rather ancient, survey, see Clark,
Covenants and Interests Running with the Land
(1947). For a more brief presentation, see Cribbet
(1962) Principles of the Law of Property, Pt. 4,
Ch. 1.

5. An equitable servitude is a legal variation on the
    theme of covenants. It applies in some cases where
    a restriction does not fulfill the requirements for
    a covenant running with the land, but would be
    given effect anyway in equity. "The doctrine is,
    in brief, that when, on a transfer of land, there
    is a covenant or even an informal contract or
    understanding that certain restrictions in the use
    of the land conveyed shall be observed, the
    restrictions will be enforced by equity, at the
    suit of the party or parties intended to be
    benefited thereby, against any subsequent owner of
    the land except a purchaser for value without
    notice of the agreement." Casner and Leach (1969)
    Property. See also, _Tulk_ v. _Moxhay_, 41 Eng. Rep.
    1143 (1848).

<u>APPENDIX A</u>

<u>RIPARIAN LAW</u>

Nature Of The Riparian Right

Riparian rights are property rights that apply to the use of water. Riparian rights are acquired by obtaining title to land that borders a natural stream, river, lake or pond.

"Riparian" comes from the Latin word "riparius," of or belonging to the bank of a river, in turn derived from "ripa," a bank, and is defined as "pertaining to or situated on the bank of a river" (78 Am. Jur. 2d S260). The riparian right to water is the right to capture and use water as contrasted with ownership of the water itself. The general rule of riparian law is that riparian owners have equal rights to use available water and can make any reasonable use of such water, so long as there is no unreasonable interference with the rights of others. Therefore, riparian rights are highly relative. If an owner of the upper part of a stream temporarily diverts substantially all of the natural water flow for use to the detriment of the owner of the lower part of the stream, for example, the latter can enjoin the first owner's use as being unreasonable. If the second owner does not suffer because of the diversion, the first owner's use of the water may be proper.

An exact determination of what constitutes a reasonable use depends on the circumstances confronting all the owners and an evaluation of conflicting interests. The Supreme Court of North Carolina has stated that reasonable use "is a question of fact having regard to the subject-matter and the use; the occasion and manner of its application; its object and extent and necessity; the nature and the size of the stream; the kind of business to which it is subservient; the importance and necessity of the use claimed by one party, and the extent of the injury caused by it to the other" (<u>Dunlap</u> v. <u>Carolina Power & Light Co.</u> 1938 212 N.C. 814, 820, 195 S.E. 43, 47).

Traditionally, domestic uses (washing, drinking, cooking, watering livestock, etc.) have been reasonable per se. However, artificial uses (irrigation, manufacturing, power, recreation, etc.) have more commonly been subject to dispute. Although water may be diverted at any point on riparian land, use outside the riparian parcel is "unreasonable" with little exception (78 Am. Jur 2d S290). The reason for restricting water use to riparian land has purportedly been the fear that riparian owners would suffer a shortage of water if non-owners were allowed access (Lauer 1970).

In times of water shortages, riparian law requires that domestic uses take priority over artificial uses and that the latter uses be apportioned on a "reasonable" basis.

Riparian rights are not lost by non-use or in any way diminished with respect to other riparians because of subsequent ownership. However, riparian rights may be lost by condemnation, prescription (adverse use by an upstream user for a specified period of time), changing stream channel location, and transfer of water use rights without conveyance of riparian land.

Short of actual loss, a diminution in relative value of riparian rights may result from state and federal regulation for health, safety, or enviromental purposes.

Prevalence of Riparian Law

Although the riparian law doctrine is recognized in most state jurisdictions, the extent of recognition varies greatly from one state to the next. It is generally accepted that riparian law was introduced to the common law of the United States in the 19th century and derived from the common law of England. The eastern states recognized the doctrine initially. As the population moved westward, the doctrine was also accepted by the western states and territories. Over time, the riparian system has proved unsatisfactory in the more arid western states as contrasted with the eastern states where water is more plentiful. The shortcomings of the riparian system in the west are attributable to a number of factors: (1) the system prohibited diversion of water to non-riparian land; (2) the riparian holder was entitled to make inefficient use of water; and (3) the system failed to provide adequate guidelines for the use of ground water and diffuse surface water.

In view of the inadequacy of the doctrine in the west, it is no longer recognized in at least eight states: Arizona, Colorado, Idaho, Montana, Nevada, New Mexico, Utah, and Wyoming (Hutchins 1971). These states have adopted the appropriation doctrine of water rights in lieu of the riparian system. Briefly stated, the appropriation doctrine grants a prior user of water superior rights over a subsequent user so long as the prior user continues to apply the water for a beneficial use.

The riparian system is partially recognized in Alaska, California, Kansas, Nebraska, North Dakota, Oklahoma, Oregon, South Dakota, Texas, and Washington (Hutchins 1971).

The degree of riparian recognition in the above states depends on state case law and legislative enactment. Generally, recognition is curtailed by coexistence of the appropriation doctrine of water rights. In such instances, riparian owners are accorded varying forms of "grandfather rights." For example, in Kansas, common law riparian rights are limited to the use of water actually applied to a beneficial use at the time of enactment of the reform legislation or within a reasonable time thereafter (Kansas Laws 1945, Chpt. 390; Laws 1957, Chpt. 539; Stat. Ann. S82a-701 et seq. 1969). A law passed in the state of Washington in 1967 provides that a riparian owner who does not exercise riparian rights for any period of five successive years after the effective date of the law shall relinquish such rights or portions thereof, and the affected water shall become available for appropriation (Laws 1967, Chpt. 233; Rev. Code S90.14.170 Supp. 1970).

The riparian doctrine has been altered in the eastern states of Maryland, Minnesota, Florida, and Iowa by a permit system (Md. Code Ann. Art. 96A, S11 Supp. 1969; Minn. Stat. Ann. S105.41 Supp. 1969; Fla. Stat. Ann. S 373.100 1960; Iowa Code Ann. S455a. 26 Supp. 1969). However, the riparian system still exists in these states for domestic water uses. The state of Mississippi has also altered the riparian doctrine by adopting the appropriation doctrine and allowing riparians priority to perfect their rights through new statutory procedures (Miss. Code Ann. S5956-04 Supp. 1968).

Aquaculture And The Riparian System

The riparian system results in uncertainty as to what uses of water are permissible (or reasonable) and as to how much water may be used. The development of aquaculture, as a use of water, may be adversely affected in view of such uncertainty. As one author has stated about the riparian doctrine: "There is insecurity in any investment for developing, storing, or using water, because there is no assurance that other landowners will not some day undertake to develop and use the same water" (Thomas 1958). Thus, existing riparian owners, and even future riparian owners, may at any time initiate a new use or enlarge upon an

existing one to the detriment of other owners. The net
effect is curtailment of future beneficial uses. Stated by
the United States Supreme Court: "Riparianism, pressed to
the limits of its logic, enables one to play dog-in-the-
manger" (U.S. v. Gerlach Livestock Co. 1950, 339 US 725,
751, 70S. Ct. 955, 969).

The ultimate determination of the reasonable use
question depends on conflict between competing uses. Only
upon the occurrence of such a controversy will a court
entertain a legal action. Although a court may resolve the
reasonable use issue between the litigants, such a
determination may not be reasonable against current and
future riparian owners who are not parties to the action.

Generally, courts have refused to consider the following
factors in determining reasonableness of use: (1) priority
of use; (2) non-use; (3) extent of riparian frontage or
riparian land; and (4) preference between classes or types
of use other than domestic uses (Lauer 1970).

Development of any form of aquaculture beyond the
present state-of-the-art unquestionably depends on the
greater use of water as it naturally occurs or by diversion
to artificial retention areas. Greater use of water in the
United States mandates more efficient use. The repudiation
and modification by the western states of the riparian
doctrine serves as nearly uncontrovertible evidence of the
undesirability of riparian law as an answer to the best
maximum use of water. As demand for water in the eastern
U.S. increases, forces for reform of the riparian doctrine
of water law may also increase.

REFERENCES

Hutchins, W.A. (1971) Water Rights in the Nineteen Western
  States. Economic Research Service, U.S. Department of
  Agriculture. Vol. I:206-225. Washington, D.C.: U.S.
  Department of Agriculture.

Lauer, T.E. (1970) Reflections on Riparianism. Missouri Law
  Review 15:1, 5.

Thomas, H.E. (1958) Hydrology v. water allocation in the
  eastern United States. Pages 165-180, in The Law of
  Water Allocation in the Eastern United States, edited by
  D. Haber and S. Bergen. New York: Ronald.

<u>APPENDIX B</u>

<u>CURRENT STATUS OF GENETICS AND SELECTIVE BREEDING</u>
<u>IN MAJOR AQUACULTURE SPECIES</u>

Information on the status of genetics and selective breeding
in several important species is presented here in slightly
more technical terms for readers with specific interests in
these species.

Common Carp

Intra-specific hybridization has been responsible for the
development of many commercial strains of the common carp,
particularly in the U.S.S.R. Favorable qualities of
different populations, such as tolerance for low temperature
and rapid rate of growth, have been successfully combined by
such breeding.

Experience has shown that the common carp of European
origin has apparently reached a selection plateau for growth
rate and, therefore, further intra-line selection for this
character yields only negligible results. On the other
hand, heterosis in F_1 crossbreds resulted in improved vigor,
growth capability, viability and fertility. Crossbreeding
has already been adopted commercially in some carp-raising
countries, such as Israel, the U.S.S.R and Yugoslavia. The
use of a small number of breeding stocks has frequently
resulted in strong negative effects of in-breeding
depression. The solution to this problem, as indicated by
recent studies, appears to be crossbreeding. A comparison
of the "big belly" strain of carp (originally from China)
with the European carp and its crossbreds has shown that the
European carp has a much superior growth capacity, although
the "big belly" is a hardier fish with greater resistance to
diseases and to unfavorable environmental conditions. Some
of these crossbreds possess the superior growth capacity of
the European carp, but maintain much of the hardiness of the
"big belly" carp. Selection for resistance to specific
disease has also had positive results.

Reduction of intra-muscular bones in common carp will
certainly enhance its consumer acceptance. Preliminary
results of selection against such bones have been promising.

Selection for growth under intensive culture has shown
the need for further studies to develop new strains for
better adaptability to such culture conditions.

Salmonids

Progress in selective breeding of salmonids in the United
States has been achieved by selection of several species of
the family Salmonidae. Selective breeding has led to
improvements in many traits, including: growth capacity,
fecundity, proper conformation, disease resistance, age at
maturation and adaptation to new environments.

Hybridization has also given very good results and
enabled the exploitation of new ecological circumstances.
For example, the "splake," a cross between the lake trout
(_Salvelinus namaycush_) and the eastern brook trout
(_Salvelinus fontinalis_), performed successfully in the Great
Lakes where indigenous trout have been almost eliminated by
the sea lamprey. The "splake" occupies a different water
zone than the local trout, and in this way was able to
escape infestation by the lamprey. Another example is an
interracial cross between the non-migratory and the
anadromous rainbow trout, both _Salmo gairdnerii_. The
interracial hybrid possessed the good growth capacity of the
non-migratory rainbow, while preserving the migratory habit
of its second parent.

A large-scale salmonid genetics and breeding program was
established in Norway a few years ago and is suited to the
needs of that country for different stocks for commercial
expansion of the salmon industry. Genetic variability must
be maintained in stocks used in ocean ranching, and more
disease-resistant strains are needed for pen culture, where
fish are raised to market size in captivity. Some of the
commercially desirable traits of the coho, chinook, or
sockeye salmon might be acceptably combined with those of
the perch and chum salmon, which return to the sea soon
after hatching. Genetic suitability must be considered
among other factors in the transplanting of salmon stocks to
replace depleted native stocks.

Catfish

Of the many species of this group, the channel catfish
(_Ictalurus punctatus_) is the one most extensively
cultivated. The major selection work in this group has been

a choice between the available species. Intra-specific
selection in the channel catfish has been started and is now
in an early stage of experimentation. Some hybrid crosses
have given promising results. The major emphasis in
selection has been on growth rate and dressed weight
percentage. There are numerous opportunities for stock
improvement.

Tilapia

In tilapia, emphasis has been placed on all-male intra-
specific hybrids, for monosex culture. Although 100 percent
male progeny have been obtained, in no case was this a
consistent trait. This problem, therefore, requires further
research. Since the main constraint to tilapia culture in
the United States is the intolerance of this fish to low
temperatures, trial selection for tolerance to low
temperatures should have high priority.

Marine Shrimp

While pandalid shrimp females can currently be matured in
captivity, work is only beginning to progress on the sexual
maturation of captive female penaeid shrimp. Most genetic
work necessitates breeding from generation to generation and
because of poor control of reproduction, genetics of the
commercially more important penaeid shrimp is limited. Such
work will be of utmost importance in the future because
successful shrimp culture in the United States may have to
be conducted in intensive systems necessitating strains
genetically suited to intensive culture. Extensive culture
of shrimp, as presently practiced in Central America, might
benefit from trial hybridization, which uses females matured
in nature. No work has been done to delineate natural
populations, and, therefore shrimp breeders cannot be fully
aware of the genetic diversity of stocks available to them.

Freshwater Shrimp

All stocks of the freshwater shrimp (<u>Macrobrachium
rosenbergii</u>) in the United States are descendant of two
berried females. Research in Hawaii, where the species is
commercially cultured, shows the stocks there to have low
genetic variability. Clearly, stocks must be crossbred with
others to increase genetic variability before selective
breeding can be successful. Breeders would like to have
lines with improved growth rates, efficient food conversion,
shorter larval period, docility, and so on. The use of
species hybrids ought to be explored since there are more
than a dozen species of the genus <u>Macrobrachium</u> (with nearly

105

worldwide temperate and tropical distribution), including species indigenous to the U.S. mainland. There is wide range in species size.

Crayfish

Crayfish farming has depended solely on the encouragement of wild stocks. Some basic elucidation of the various stocks available and their suitability for different environments could lead to some immediate improvement.

Lobsters

The American lobster has low genetic variability, as indicated by research on iso-enzymes at the University of California. This indicates a need for hybridization to increase variability before expensive selection experiments are warranted.

Molluscs

Both gastropod and pelecypod molluscs are known to have different kinds of genetic polymorphisms. The American oyster occurs over a wide geographic range and has genetic adaptations to environmental conditions along its range; it appears to have sufficient genetic variance to respond positively to selective breeding. Selection responses have already been obtained for growth rate and disease resistance. Japanese and American oysters have been crossed within species and between species, but for the most part have not been tested adequately, and the use of hybrids has not been considered. Inter-species crosses have not been made on a large enough scale to evaluate commercial suitability. Very few genetic studies have been conducted on the European oyster.

One of the major problems in hatchery production of the American oyster is the great sensitivity of the larval stages. Unexplained larval mortality has forced the closure and relocation of hatcheries. Theoretically more sturdy larvae could be bred. Effects of inbreeding on these sensitive larvae ought to be studied.

A hybrid of the northern and southern United States hard clam was found to have some distinct advantage over non-hybrids in certain environments.

Genetic studies are currently being conducted on the bay scallop; those studies offer opportunities for self-fertilization as well as for cross-fertilization. Studies

have been published on heritability estimates, gene
frequencies, and chromosomes of the pearl oyster (which is
really a mussel). Species crosses of the abalone have been
made in Japan. Genetic variability has not been measured in
the abalone and selection experiments have not been
conducted. Slow growth of the abalone is a major deterrent
to its commercial production, and attempts at genetic
improvement of growth are warranted.

Aquatic Plants

Difficulty in control of the life cycle of seaweed has
deterred attempts at selective breeding. A Japanese worker
has made hybrids of different monoecious and dioecious types
and between monoecious and dioecious plants. Marine plants
having both sexual and asexual forms of reproduction afford
an interesting opportunity for use of induced mutations.
Once induced, a single, new, useful mutation in such species
could be propagated asexually without the random segregation
and recombination of genes that go with sexual reproduction.

REFERENCES

American Littoral Society (1972) Special News Letter Alert:
    Fin Rot.

Anonymous (1951) Diseased stripers in Connecticut are safe
    to eat. Salt Water Sportsman, June 22, 1951, page 1.

Anonymous (1972) Clams from polluted areas show measurable
    signs of stress. Maritimes 16:15-16.

Burdick, G.E. et al. (1964) The accumulation of DDT in lake
    trout and the effect on reproduction. Transactions of
    the American Fisheries Society 93:127.

Butler, P.A. (1960) Effect of pesticides on oysters. In,
    Proceedings of the National Shellfish Association 51:23.

Butler, P.A. (1969a) Monitoring pesticide pollution.
    Bioscience 19:889-891.

Butler, P.A. (1969b) The significance of DDT residues in
    estuarine fauna. Pages 205-220, Chemical Fallout, edited
    by M.W. Miller and G.G. Berg. Springfield, Ill.: C.C.
    Thomas.

Davis, H.C. (1961) Effects of some pesticides on eggs and
    larvae of oysters (_Crassostrea virginica_) and clams
    (_Venus mercenaria_). Commercial Fisheries Review
    23(12):8-23.

Davis, H.C. and H. Hidu (1969) Effects of pesticides on
    embryonic development of clams and oysters and on
    survival and growth of larvae. Fisheries Bulletin (U.S.
    Fish and Wildlife Service) 67:393-404.

Duke, T.W., J.I. Lowe, and A.J. Wilson, Jr. (1970) A
    polychlorinated biphenyl (Aroclor 1254) in the water,
    sediment, and biota of Escambia Bay, Florida. Bulletin
    of Environmental Contamination and Toxicology 5:171-180.

Farley, C.A., W.G. Banfield, G. Kasnic, Jr., and W.S. Foster
    (1972) Oyster herpes-type virus. Science 178:759-760.

Food and Agriculture Organization (1976) Technical
     Conference on Aquaculture, Tokyo. Rome: Food and
     Agriculture Organization.

Fryer, J.L., J.S. Rohovec, G.L. Tebbit, J.S. McMichael and
     K.S. Pilcher (1976) Vaccination for control of
     infectious diseases in Pacific salmon. Fish Pathology
     10:155-164.

Grossman, K. (1972) Oil spills linked to cancer. Long Island
     Press, February 5, 1972.

Halstead, B.W. (1971) Toxicity of marine organisms caused by
     pollutants. Technical Conference on Marine Pollution and
     its effects on Living Resources and Fishing. Doc.
     MP/70/R-6. Rome: Food and Agriculture Organization.

Herman, R.L. (1970) Prevention and control of fish diseases
     in hatcheries. Pages 3-15, A Symposium on Diseases of
     Fishes and Shellfishes, edited by S.F. Snieszko.
     Washington, D.C.: American Fisheries Society.

Janssen, W.A. (1970) Fish as potential vectors of human
     bacterial diseases. Pages 284-290, A Symposium on
     Diseases of Fish and Shellfish, edited by S.F. Snieszko.
     Washington, D.C.: American Fisheries Society.

Janssen, W.A. and C.D. Meyers (1968) Fish: serologic
     evidence of infection with human pathogens. Science
     159:547-548.

Johnson, S.K. (1975) Handbook of shrimp diseases.
     Publication SG-75-603. College Station, Tex.: Texas A&M
     University.

Korringa, P. (1968) Biological consequences of marine
     pollution with special reference to the North Sea
     Fisheries. Helgolander wiss. Meeresunters 17:126-140.

Krantz, G. (1970) Lymphocystis in striped bass, <u>Roccus
     saxatilis</u>, in Chesapeake Bay. Chesapeake Science 11:137-
     139.

Kuhnhold, W.W. (1971) The influence of crude oils on fish
     fry. Food and Agriculture Organization Fisheries
     Reports No. 99:157. Rome: Food and Agriculture
     Organization.

Landis, A.T. (1972) Mercury in FPC. Undersea Technology.
     December, 1972.

Mahoney, J. (1970) Special microbiological problems in
     estuarine and coastal areas. Pages 177-178, Progress in

Sport Fishery Research 1969. Resource Publication No.
88. U.S. Fish and Wildlife Service. Washington, D.C.:
Department of the Interior.

Mason, J.O. and W.R. McLean (1962) Infectious hepatitis
traced to the consumption of raw oysters. An
epidemiologic study. American Journal of Hygiene 75:90.

Metcalf, T.G. and W.C. Stiles (1966) Survival of enteric
viruses in estuary waters and shellfish. Pages 439-447,
Transmission of Viruses by the Water Route, edited by G.
Berg. New York: Wiley-Interscience.

Meyer, F.P., R.A. Schnick, K.B. Cumming, and B.L. Berger
(1976) Registration status of fishery chemicals.
Progressive Fish Culturist 38(1):3-7.

Okubo, K. and T. Okubo (1972) Study on the bioassay method
for the evaluation of water pollution, II. Use of
fertilized eggs of sea urchins and bivalves. Tokai
Regional Fisheries Research Laboratory 32:131-140.

Regier, L.W. (1972) Mercury removal from fish protein
concentrate. Journal of the Fisheries Research Board of
Canada 29:1777-1779.

Reid, J.R. (1975) Comparative efficiency of animals in the
conversion of feedstuffs to human food. Pages 16-24, In,
Proceedings, 1975 Cornell Nutrition Conferences.

Risebrough, R.W. (1969) Chemical fallout. First Rochester
Conference on Toxicity. Springfield, Ill.: C.C. Thomas.

Sidhu, G.S., G.L. Vale, J. Shipton, and K.E. Murray (1971)
Nature and effects of a kerosene-like taint in mullet
(Mugil cephalus). Fisheries Reports 99:143. Rome: Food
and Agriculture Organization.

Sindermann, C.J. (1970) The role of control of diseases and
parasites in mariculture. Pages 145-173. In,
Proceedings, Second Conference on Food-Drugs from the
Sea, edited by H.W. Youngken, Jr. Washington, D.C.:
Marine Technology Society.

Sindermann, C.J. (1972) Some biological indicators of marine
environmental degradation. Journal of the Washington
Academy of Science 62:184-189.

Sindermann, C.J., ed. (1974) Diagnosis and control of
mariculture diseases in the United States. Middle
Atlantic Coastal Fisheries Center, National Marine
Fisheries Service, Technical Service Report No. 2.

Sindermann, C.J. (1976) Oyster mortalities and their
     control. Food and Agriculture Organization Technical
     Conference on Aquaculture, Tokyo. Rome: Food and
     Agriculture Organization.

Smith, R.R. (1971) A method for determining digestibility
     and metabolizable energy of fish feeds. Progressive Fish
     Culturist 33:132-134.

Smith, R.R. (1976) Studies on the energy metabolism of
     cultured fishes. Ph.D. Thesis. Ithaca, N.Y.: Cornell
     University.

U.S. Congress, House (1976) Conference Report to Accompany
     HR 200. House Report 94-711, 94th Congress, 2nd Session.

Wisely, B. and R.A.P. Blick (1967) Mortality of marine
     invertebrate larvae in mercury, copper, and zinc
     solutions. Australian Journal of Marine and Freshwater
     Research 18:63-72.

Woelke, C.E. (1967) Measurement of water quality with the
     Pacific oyster embryo bioassay. Pages 112-120, Water
     Quality Criteria, ASTM, STP 416, American Society of
     Testing Materials.

Woelke, C.E. (1968) Application of shellfish bioassay
     results to the Puget Sound pulp mill pollution problem.
     Northwest Science 42:125-133.

BIBLIOGRAPHY

Acton, R.T., P.W. Weinheimer, and E.E. Evans (1969) A
    bactericidal system in the American lobster, Homarus
    americanus. Journal of Invertebrate Patholology 13:463-
    464.

Anderson, E. (1949) Introgressive hybridization. Facsimile
    of the 1949 ed. New York: Hafner Publishing Company.

Anderson, L.G. (1977) The Economics of Fisheries Management.
    Baltimore, Md.: The Johns Hopkins University Press.

Anderson, L.G. and D.C. Tabb (1971) Some economic aspects of
    pink shrimp farming in Florida. In, Proceedings, Gulf
    and Caribbean Fisheries Institute 23 (1970):113-124.
    Coral Gables, Fla.: University of Miami.

Andrews, J.D. and M. Frierman (1974) Epizootiology of
    Minchinia nelsoni in susceptible wild oysters in
    Virginia, 1959 to 1971. Journal of Invertebrate
    Patholology 24:127-140.

Anonymous (1976) Apalachicola--our last clean major river
    endangered by Corps of Engineers development plan. ENFO
    Florida Report. Winter Park, Fla.: Florida Conservation
    Foundation, Inc.

Aquatic Food Animal Task Force (1975) Freshwater foodfish
    (and crawfish)-- Research Needs in the Southern Region.
    Research Publication 407. Agricultural Experiment
    Station, Auburn, Ala.: Auburn University.

Avala, F.J. (1968) Genotype, environment and population
    members. Science 162:1453-1459.

Avault, J.W., Jr. (1972) Crayfish farming in the United
    States. Paper presented at the First International
    Symposium on Freshwater Crayfish, September 12-15, 1972,
    Austria. Sea Grant Preprint 897, 23 pp. Center for
    Wetland Resources, Baton Rouge, La.: Lousiana State
    University.

Bailey, W.M., M.D. Gibson, S.H. Newton, J.M. Martin, and
     D.L. Gray (1976) Status of Commercial aquaculture in
     1975. In, 30th Annual Conference of the Southeastern
     Association of Game and Fish Commissioners, Jackson,
     Mississippi.

Ball, R.C., and E.H. Bacon (1954) Use of pituitary material
     in the propagation of minnows. Progressive Fish
     Culturist 16:108-113.

Bardach, J.E., J.H. Ryther, and W.O. MacLarney (1972)
     Aquaculture, the farming and husbandry of freshwater and
     marine organisms. New York: John Wiley and Sons, Inc.
     868 p.

Bell, F.W. and E.R. Canterberry (1976) Aquaculture for
     Developing Countries: A Feasibility Study. Cambridge,
     Massachusetts: Ballinger Publishing Company.

Bente, P.F., Jr. (1971) Application of the law. In,
     Proceedings, First American Workshop. Pages 50-54, World
     Mariculture Society (1970). Baton Rouge, La.: Louisiana
     State University.

Blaxter, J.H.S., Ed. (1973) The Early Life History of
     Fishes. New York: Springer-Verlage.

Boney, A.D. (1965) Aspects of the biology of the seaweeds of
     economic importance. Advances in Marine Biology (3) 105-
     253.

Bonn, E.W., W.M. Bailey, J.D. Bayless, T.E. Erickson, and
     R.E. Stevens (1976) Guidelines for striped bass culture.
     Striped Bass Committee of the Southern Division of the
     American Fisheries Society.

Brett, J.R. (1976) Scope for metabolism and growth of
     sockeye salmon (Oncorhynchus nerka) and some related
     energetics. Journal of the Fisheries Research Board of
     Canada 33:307-313.

Charbonneau, J.J. and R. Marasco (1975) A positive spatial
     equilibrium model of oyster markets: a simultaneous
     equations approach. Agricultural Experiment Station
     College Park, Md.: University of Maryland.

Cherfas, B.I., ed. (1969) Genetics, Selection, and
     Hybridization of Fish. National Marine Fisheries
     Service, National Oceanic and Atmospheric
     Administration, U.S. Department of Commerce and the
     National Science Foundation. Washington, D.C.: Israel
     Program for Scientific Translations. (Translated from
     Russian 1972).

Clemens, H.P. and K.E. Sneed (1957) The spawning behavior of
    the channel catfish, _Ictalurus punctatus_. Special
    Science Report, Fish 219, U.S. Fish and Wildlife
    Service, Washington, D.C.: U.S. Department of the
    Interior.

Clemens, H.P. and K.E. Sneed (1962) Bioassay and use of
    pituitary materials to spawn warm-water fishes. U.S.
    Fish and Wildlife Service, Washington, D.C.: U.S.
    Department of the Interior.

Combs, B.D. and R.E. Burrows (1959) Effects of injected
    gonadotrophins on maturation and spawning of blueback
    salmon. Progressive Fish Culturist 21(4):164-171.

Cowey, C.B. and J.R. Sargent (1976) Fish nutrition. Advanced
    Marine Biology 10:383-492.

Crutchfield, J.A. and G. Pontecorvo (1969) The Pacific
    Salmon Fisheries. A Study of Irrational Conservation.
    Resources of the Future, Inc. Baltimore: Johns Hopkins
    Press.

Davidson, J.R. (1972) Economics of aquaculture development.
    _In_, Proceedings, Fourth National Sea Grant Conference,
    Oct. 12-13, 1972. Sea Grant Publ. WIS-SG-72-112. Sea
    Grant Communications Office: Madison, Wisc.: University
    of Wisconsin.

Davies, C.B. and H.W. Shields (1971) Mariculture and the
    law. _In_, Proceedings, First Annual Workshop, World
    Mariculture Society (1970). Baton Rouge, La.: Louisiana
    State University.

DeBoer, J.D. and J.H. Ryther (1977) Preliminary studies on a
    combined seaweed mariculture-tertiary waste treatment
    system. _In_, Costa Rica Proceedings World Mariculture
    Society, January, 1977.

de la Bretonne, L., Jr. (1977) A review of crayfish culture
    in Louisiana. Eighth Annual Meeting of the World
    Mariculture Society. San Jose, Costa Rica.

Dillon, O.W., Jr. (1976) Aquaculture in the Southern United
    States. Food and Agriculture Organization Technical
    Conference on Aquaculture, Kyoto, Japan.
    FIR:AQ/Conf./76/E.26. Rome: Food and Agriculture
    Organization.

Dobie, J., O.L. MeeLean, S.F. Snieszco, and G.N. Washburn
    (1956) Raising bait fishes. U.S. Fish and Wildlife
    Service, U.S. Department of the Interior. Washington,
    D.C.

Dobzhansky, T. (1951) Genetics and the Origin of Species, 3rd edition. New York: Columbia University Press.

Doryshev, S. (1973) Aquaculture in the Soviet Union. In, Marine Fisheries Review. 35(10):42-43. National Marine Fisheries Service, National Oceanic and Atmospheric Administration, Washington, D.C.: U.S. Department of Commerce.

Fontenele, O. (1955) Injecting pituitary (hyphyseal) hormones into fish to induce spawning. Progressive Fish Culturist 17:71-75.

Food and Agriculture Organization (1972) Report of the first meeting of the FAO ad hoc working party on genetic resources of fish. December 7-10. Fisheries Reports No. 19. Rome: Food and Agriculture Organization.

Food and Agriculture Organization (1976) Summary Report of the Technical Conference on Aquaculture, Kyoto, Japan. Rome: Food and Agriculture Organization.

Food and Agriculture Organization (1977) A Selected Bibliography on the economic aspects of aquaculture, 1969 to 1977. Food and Agriculture Organization Fisheries Circular No. 702. Rome: Food and Agriculture Organization.

Galtsoff, P.S. (1964) The American oyster. Fishery Bulletin 64:1-480. U.S. Fish and Wildlife Service, U.S. Department of the Interior.

Gaucher, T.A., ed. (1971) Aquaculture: A New England Perspective. The New England Marine Resources Informatin Program. Narragansett, R.I.: University of Rhode Island.

Goldman, J.C. and J.H. Ryther (1975) Nutrient transformations in mass cultures of marine algae. Journal of Environmental Engineering 101:351-364.

Gowen, J.W. (1964) Heterosis. A Record of Researches Directed toward Explaining and Utilizing the Vigor of Hybrids. New York: Hafner Publishing Company.

Halver, J.E., ed. (1972) Fish Nutrition. New York: Academic Press.

Hanamura, N., T. Ishida, S. Sano, W.E. Ricker, and F. Neave (1966) Salmon of the North Pacific Ocean. Part III. A review of the life history of North Pacific salmon. Bulletin 18. International North Pacific Fish Commission.

Hasler, A.D. (1966) Underwater Guideposts, Homing of Salmon.
     Madison, Wisc.: University of Wisconsin Press.

Hasler, A.D., R.K. Meyer, and H.M. Field (1939) Spawning
     induced prematurely in trout with the aid of pituitary
     glands of the carp. Endocrinology 25:978-983.

Hendry, R.R. (1971) The Florida Mariculture law. In,
     Proceedings, First Annual Workshop, World Mariculture
     Society. Baton Rouge, La.:Louisiana State University.

Hinshaw, R.N. (1973) Pollution as a result of fish cultural
     activities. Office of Environmental Monitoring, U.S.
     Environmental Protection Agency. Washington, D.C.: U.S.
     Environmental Protection Agency.

Idyll, C.P., D.C. Tabb, W.T. Yang, and E.S. Iversen (1969)
     Shrimp and Pompano Culture Facilities at the University
     of Miami. Sea Grant Information Bulletin 2. Coral
     Gables, Fla.: University of Miami.

International Conference on Aquaculture Nutrition (1976).
     Newark, N.J.: University of Delaware Press.

International Congress on Nutrition (1975) Nutrition and
     Production of Fish. Proceedings, Ninth International
     Congress. S. Karger and Co., Basel. 3:142-220.

Johnston, R.S. (1976) The Relationship between Property
     Rights Arrangements and the Nature of Aquacultural
     Developments. Discussion paper presented at the Small
     Scale Fisheries Development Planning Meeting, September
     1976. Honolulu, Ha.: East-West Center. (Unpublished).

Johnston, W.E. and D.W. Collinsworth (1973) An Annotated
     Bibliography for Economic Evaluations of the Aquaculture
     of Selected Crustaceans and Mollusks. Sea Grant
     Publication No. 2. San Diego, Calif.: University of
     California.

Johnston, W.E. and P.G. Allen (1974) Technology Assessment
     and the Direction for Future Research: Use of
     Computerized Budgeting for Lobster Aquaculture. Paper
     presented at the annual meeting of the American
     Agricultural Economics Association, College Station,
     Texas.

Kane, T.E. (1969) Aquaculture and the law. Thesis in Ocean
     Law. Coral Gables, Fla.: University of Miami.

Ketchum, B., ed. (1972) Aquaculture in the coastal zone.
     Pages 48-51, The Waters Edge: Critical Problems of the
     Coastal Zone. Cambridge, Mass.: MIT Press.

Ketola, H.G. (1976) Quantitative Nutritional Requirements of
    Fishes for Vitamins and Minerals. Feedstuffs 2(16):42-
    44.

Klontz, G.W. and J.G. King (1975) Aquaculture in Idaho and
    nationwide. Boise, Idaho: Idaho Department of Water
    Resources.

Kolby, A.C., Jr. (1972) Food and Drug Administration
    guidelines for contaminants in fishery products. In,
    Proceedings, Gulf and Caribbean Fisheries Institute.
    Coral Gables, Fla.: University of Miami.

Landy, B.A. (1975) Constraints on aquaculture projects.
    Marine Fisheries Review 37(1):33-35.

Lapointe, B.E., L.D. Williams, J.C. Goldman, and J.H. Ryther
    (1976) The mass outdoor culture of macroscopic marine
    algae. Aquaculture 8:9-21.

Lerner, I.M. and H.P. Donald (1966) Modern Developments in
    Animal Breeding. New York: Academic Press.

Ling, S.W. and T.J. Costello (1976) Review of the culture of
    freshwater prawns. Food and Agriculture Organization
    Technical Conference on Aquaculture. Kyoto.
    FIR:AQ/Conf./76/R.29. Rome: Food and Agriculture
    Organization.

Lockwood, G.S. (1977) An Analysis of Constraints and
    Stimulants to Aquaculture Development in the United
    States. Monterey, Calif.: Monterey Abalone Farms.

Longwell, A.C. (1974) Some impressions regarding genetics
    and the fisheries of Japan. In. Proceedings. First U.S.-
    Japan Meeting on Aquaculture, Tokyo, Japan. National
    Marine Fisheries Service Circular 388. Washington, D.C.:
    U.S. Department of Commerce.

Longwell, A.C. (1976) Review of genetic and related studies
    on commercial oysters and other pelecypod mollusks.
    Journal of the Fisheries Research Board of Canada.
    Special Issue 33(4):1100-1107.

Longwell, A.C. and S.S. Stiles (1973) Oyster genetics and
    the probable future role of genetics in aquaculture.
    Malacological Review 6:151-177.

Loosanoff, V.L. and H.C. Davis (1963) Rearing of bivalve
    mollusks. Advances in Marine Biology, Vol. 1. edited by
    F.S. Russell. London: Academic Press. pp. 1-136.

Lutz, R.A. (1977) A comprehensive review of commercial mussel industries in the United States. Report No. 76-25. Walpole: University of Maine.

Mardela Corporation (1973) Summary report to participants, NOAA regional aquaculture workshop. Burlingame, California.

McCoy, E.W. and A.B. Sherling (1973) Economic analysis of the catfish processing industry. Agricultural Experiment Station, circular 207; Auburn, Ala.: Auburn University. 20 pp.

McNeil, W.J., ed. (1970) Marine Aquaculture. Corvallis, Oreg.: Oregon State University Press.

McNeil, W.J. and J.E. Bailey (1975) Salmon Rancher's Manual. Northwest Fisheries Center, National Oceanic and Atmospheric Administration, U.S. Department of Commerce.

Merrill, A.S. and H.S. Tubiash (1970) Molluscan resources of the Atlantic and Gulf Coast of the United States. Pages 925-948. In, Proceedings, Symposium on Mollusca--Part III.

Meryman, H.T. (1966) Cryobiology. New York: Academic Press.

Meyer, F.P., K.E. Sneed, and P.T. Eschmeyer, eds. (1973) Second report to Fish Farmers--The status of warmwater fish farming and progress in fish farming research. Resource Publication 113. Bureau of Sport Fisheries and Wildlife. U.S. Fish and Wildlife Service, Washington, D.C.: U.S. Department of the Interior.

Miller, M.M. and D.A. Nash (1969) The development of catfish as a farm crop and an estimation of its economic adatability to radiation processing. National Marine Fisheries Service, Washington, D.C.: U.S. Department of Commerce. 136 pp.

Milne, P.H. (1972) Fish and Shellfish Farming in Coastal Waters. Fishing News (Books) Ltd.

Muntzing, A. (1963) Origin and Breeding of our Cultivated Plants. Kungl. Skogs- Och Lantbruksakademiens Tidskrift.

National Oceanic and Atmospheric Administration (1977) NOAA Aquaculture Plan. National Marine Fisheries Service, National Oceanic and Atmospheric Administration. Washington, D.C.: U.S. Department of Commerce.

National Research Council (1961) Mutation and Plant
Breeding. Agricultural Board. Washington, D.C.: National
Academy of Sciences.

National Research Council (1972) Genetic Vulnerability of
Major Crops. Board on Agriculture and Renewable
Resources, Commission on Natural Resources. Washington,
D.C.: National Academy of Sciences.

National Research Council (1974) Nutrient Requirements of
Trout, Salmon and Catfish. Board on Agriculture and
Renewable Resources, Commission on Natural Resources.
Washington, D.C.: National Academy of Sciences.

National Research Council (1974) Research Needs in Animal
Nutrition. Board on Agriculture and Renewable Resources,
Commission on Natural Resources. Washington, D.C.:
National Academy of Sciences.

National Research Council (1977) World Food and Nutrition
Study: Panel on Aquatic Food Resources, Commission on
International Relations. Washington, D.C.: National
Academy of Sciences.

National Research Council (1977) Nutrient Requirements of
Warmwater Fishes. Board on Agriculture and Renewable
Resources, Commission on Natural Resources. Washington,
D.C.: National Academy of Sciences.

Neuhaus, O.W. and J.E. Halver (1969) Fish in Research. New
York: Academic Press.

Nitta, T. (1971) Marine Pollution in Japan. Fisheries
Reports No. 99:110-111. Rome: Food and Agriculture
Organization.

Nose, T. (1971) Determination of nutritive value of food
protein in fish. III. Nutritive values of casein, white
fish meal, and soybean meal in rainbow trout
fingerlings. Bulletin Freshwater Fish Res. Lab., (Tokyo)
21:85-98.

Odum, W.E. (1973) The potential of pollutants to adversely
affect aquaculture. In, Proceedings, Gulf and Caribbean
Fisheries Institute 25. Nov. 1972. Coral Gables, Fla.:
University of Miami.

Ogino, C. and M.S. Chen (1973) Protein nutrition in fish.
IV. Biological value of dietary proteins in carp.
Bulletin Japan Society Scientific Fish. 39:797-800.

Oswald, W.J. and H.B. Gotaas (1957) Photosynthesis in sewage treatment. In, Transactions, American Society of Civil Engineers 122:73-105.

Pacific Congress (1976) Aquaculture Symposium. In, Proceedings, 13th Pacific Congress. Journal of the Fisheries Research Board of Canada 33:875-1119.

Paterson, W.D. and J.E. Stewart (1974) In vitro phagocytosis by hemocytes of the American lobster (Homarus americanus). Journal of the Fisheries Research Board of Canada 31:1051-1056.

Pickford, G.E. and J.W. Atz (1957) The physiology of the pituitary gland of fishes. New York Zoological Society, New York. Ann Arbor, Mich.: Cushing-Mallory, Inc.

Pillay, T.V.R. (1976) The state of aquaculture. The Commercial Fish Farmer and Aquaculture News 1(5):9-11.

Prather, E.E., J.R. Fielding, M.C. Johnson, and H.S. Swingle (1953) Production of bait minnows in the Southeast. Agricultural Experiment Station, Circular No. 112. Alabama Polytechnic Institute

Pritchard G.I. (1976) Structured aquaculture development with a Canadian perspective. Journal of the Fisheries Research Board of Canada 33:856-870.

Provenzano, A.J. (1971) Research priorities for non-penaeid crustaceans. In, Proceedings, Second Annual Workshop, World Mariculture Society, January 28-29, 1971. Baton Rouge, La.: Louisiana State University.

Purdom, C.E. (1972) Genetics and Fish Farming. Great Britain Ministry of Agriculture. Laboratory Leaflet.

Queirolo, L.E. (1977) Substitutional relationships between rainbow trout and pansize salmon: a market demand analysis. Unpublished Masters Thesis, Corvallis, Oreg.: Oregon State University.

Ritchie, T.P. (1977) A Comprehensive review of commercial Clam industries in the United States. Lewes, Del.: University of Delaware.

Ryther, J.H., W.M. Dunstan, K.R. Tenore, and J.E. Huguenin (1972) Controlled eutrophication--increasing food production from the sea by recycling human wastes. Bio-Science 22:144-152.

Ryther, J.H., L.D. Williams, D.C. Kneale, B.E. Lapointe, J.B. Loftus, and R.W. Stenberg (1976) Harbor Branch

Foundation Aquaculture Project, Annual Progress Report. Ft. Pierce, Florida.

Schroder, J.H. (1973) Genetics and Mutagenesis of Fish. New York: Springer-Verlag.

Schwanitz, F. (1966) The Origin of Cultivated Plants. Cambridge, Mass.: Harvard University Press.

Scott, A. (1970) Economic obstacles to marine development. In, Conference on Marine Aquiculture, edited by W.J. McNeil. Oregon State University Marine Science Center. Corvallis, Oreg.: Oregon State University Press. pp. 153-167.

Shields, H.W. (1970) Aquaculture and the law. In, Proceedings, Gulf and Caribbean Fisheries Institute 22. Coral Gables, Fla.: University of Miami.

Shultz, F.T. (1970) Genetic potentials in aquaculture. In, Proceedings, Second Conference on Food-Drugs from the Sea, edited by H.W. Youngken, Jr. Washington, D.C.: Marine Technology Society.

Simon, R.C. and P.A. Larkin, eds. (1972) The stock concept in Pacific Salmon. H.R. MacMillan Lectures in Fisheries. Vancouver, British Columbia: University of British Columbia.

Smith, A.U. (1970) Current Trends in Cryobiology. New York: Plenum Press.

Smith, F.J. and K.J. Roberts, eds. (1976) Aquaculture Economics Research Needs: Report from a Workshop to Identify Aquaculture Economics Research Needs, Atlanta, Georgia, April 23, 1976. South Carolina Sea Grant Technical Report Number 5. Charleston, S.C.: South Carolina Marine Resources Center.

Sneed, K.E. and H.P. Clemens (1959) The use of human chorionic gonadotropin to spawn warm-water fishes. Progressive Fish Culturist 21(3):117-120.

Sneed, K.E. and H.P. Clemens (1960) Hormone spawning of warm-water fishes: its practical and biological significance. Progressive Fish Culturist 22(3):109-113.

Stevens, R.E. (1964) A final report on the use of hormones to ovulate striped bass, _Roccus saxatilis_ (Walbaum). In, Proceedings, Eighteenth Annual Conference, Southeastern Association of Game and Fish Commissioners, Clearwater, Florida. Columbia, S.C.: Southeastern Association of Game and Fish Commissioners.

Stevens, R.E. (1966) Hormone-induced spawning of striped
    bass for reservoir stocking. Progressive Fish Culturist
    28(1):19-28.

Stevens, R.E. (1970) Hormonal relationship affecting
    maturation and ovulation in largemouth bass (Micropterus
    salmoides Lacepede). Unpublished Ph.D. dissertation.
    Department of Zoology, Raleigh, N.C.: North Carolina
    State University.

Stevens, R.E. (1975) Current and future considerations
    concerning striped bass culture and management. In,
    Proceedings, 28th Annual Conference of the Southeast
    Association of Game and Fish Commissioners. White
    Sulphur Springs, W.V.

Surma, M.A. and J.A. Koburger (1972) Microbial survey of
    imported shrimp. In, Proceedings, Gulf and Caribbean
    Fisheries Institute 24. Coral Gables, Fla.: University
    of Miami.

Tabb, D.C., W.T., Yang, Y. Hirono, and J. Heinen (1972) A
    manual for culture of pink shrimp, Penaeus duorarum,
    from eggs to postlarvae suitable for stocking. Sea Grant
    Special Bulletin. Coral Gables, Fla.: University of
    Miami.

Tobolski, J.J. (1977) An evaluation of private salmon
    aquaculture companies in the states of Alaska, Oregon,
    and Washington. Unpublished Thesis, Seattle, Wa.

U.S. Congress, Senate (1974) The Economic Value of Ocean
    Resources to the United States. Committee on Commerce,
    93rd Congress, 2nd Session.

Van Sickle, V.R., B.B. Barrett, T.B. Ford, and L.J. Gulick
    (1976) Barataria Basin: Salinity changes and oyster
    distribution. WLFC Technical Bulletin. 20 and Sea Grant
    Publication No. LSU-T-76-002, Center for Wetland
    Resources. Baton Rouge, La.: Louisiana State University.

Vondruska, J. (1976) Aquacultural Economics Bibliography.
    Technical Report SSRF-703. National Oceanic and
    Atmospheric Administration, National Marine Fisheries
    Service, Washington, D.C.: U.S. Department of Commerce.

Wada, K. (1977) Publications on the genetics of the Japanese
    pearl oyster. Available from the National Pearl Research
    Laboratory. Koshikojima, Mie, Japan. Milford Laboratory,
    National Marine Fishers Service.

Wahle, R.J., R.R. Vreeland, and R.H. Lander (1974)
    Bioeconomic contribution of Columbia River Hatchery coho

salmon, 1965 and 1966 broods, to the Pacific salmon
fisheries. Fishery Bulletin 72(1):139-169. National
Marine Fisheries Services, Washington, D.C.: U.S.
Department of Commerce.

Webber, H.H. (1968) Mariculture. Bioscience 18(10)940-45.

Webber, H.H. (1972) The design of an aquaculture enterprise.
In, Proceedings, the Gulf and Caribbean Fisheries
Institute 24:117-125. Nov. 1971. Coral Gables, Fla.:
University of Miami.

Wickens, J.F. (1976) Prawn biology and culture. Oceanogr.
Marine Biology Annual Review 14: 435-507.

Wolverton, B.C. and R.C. McDonald (1975) Water hyacinths for
upgrading sewage lagoons to meet advanced wastewater
treatment standard: Part I. NASA Technical Memorandum
TM-Y-72729.

Wood, J.W. (1974) Diseases of Pacific Salmon, Their
Prevention and Treatment. 2nd edition. Washington
Department of Fisheries.

Date Due

Due	Returned	Due	Returned
MAR 1 3 1980	MAR 1 1 1980		
AUG 2 8 1980	AUG 2 8 1980		
DEC 1 6 1980	NOV 2 0 1980		
SEP 2 4 1981	SEP 2 3 1981		
DEC 2 1 1981	DEC 0 9 1981		
MAR 2 5 1982	MAR 1 8 1982		
MAY 1 0 1982	MAY 1 0 1982		
FEB 0 3 1986	FEB 0 3 1986		
APR 0 7 1986	APR 0 6 1986		
OCT 2 3 1986			
AUG 2 2 1988	DEC 1 5 1988		
NOV 2 1 1988	DEC 1 5 1988		
OCT 1 2 1989	SEP 0 7 1989		
SEP 3 0 1992			
NOV 2 5 1991	OCT 2 8 1992		
OCT 2 3 1997	NOV 1 0 1997		
DEC 1 4 2005	DEC 0 7 2005		